Lecture Notes in Computer Science 15198

Founding Editors

Gerhard Goos
Juris Hartmanis

The series Lecture Notes in Computer Science (LNCS), including its subseries Lecture Notes in Artificial Intelligence (LNAI) and Lecture Notes in Bioinformatics (LNBI), has established itself as a medium for the publication of new developments in computer science and information technology research, teaching, and education.

LNCS enjoys close cooperation with the computer science R & D community, the series counts many renowned academics among its volume editors and paper authors, and collaborates with prestigious societies. Its mission is to serve this international community by providing an invaluable service, mainly focused on the publication of conference and workshop proceedings and postproceedings. LNCS commenced publication in 1973.

Esther Puyol-Antón · Ghada Zamzmi ·
Aasa Feragen · Andrew P. King ·
Veronika Cheplygina ·
Melanie Ganz-Benjaminsen · Enzo Ferrante ·
Ben Glocker · Eike Petersen · John S. H. Baxter ·
Islem Rekik · Roy Eagleson
Editors

Ethics and Fairness
in Medical Imaging

Second International Workshop on
Fairness of AI in Medical Imaging, FAIMI 2024
and Third International Workshop on
Ethical and Philosophical Issues in Medical Imaging, EPIMI 2024
Held in Conjunction with MICCAI 2024
Marrakesh, Morocco, October 6–10, 2024
Proceedings

 Springer

Editors

Esther Puyol-Antón ⓘ
King's College London
London, UK

Aasa Feragen ⓘ
Technical University of Denmark
Kgs Lyngby, Denmark

Veronika Cheplygina ⓘ
University of Copenhagen
Copenhagen, Denmark

Enzo Ferrante ⓘ
National University of the Litoral
Santa Fe, Argentina

Eike Petersen ⓘ
Technical University of Denmark
Kgs Lyngby, Denmark

Islem Rekik ⓘ
Imperial College London
London, UK

Ghada Zamzmi ⓘ
Food and Drug Administration (FDA)
Silver Spring, MD, USA

Andrew P. King ⓘ
King's College London
London, UK

Melanie Ganz-Benjaminsen ⓘ
University of Copenhagen
Copenhagen, Denmark

Ben Glocker ⓘ
Imperial College London
London, UK

John S. H. Baxter ⓘ
Université de Rennes
Rennes, France

Roy Eagleson ⓘ
Western University
London, ON, Canada

ISSN 0302-9743   ISSN 1611-3349 (electronic)
Lecture Notes in Computer Science
ISBN 978-3-031-72786-3   ISBN 978-3-031-72787-0 (eBook)
https://doi.org/10.1007/978-3-031-72787-0

This Springer imprint is published by the registered company Springer Nature Switzerland AG
The registered company address is: Gewerbestrasse 11, 6330 Cham, Switzerland

If disposing of this product, please recycle the paper.

FAIMI Preface

In recent years, research into fairness in machine learning has highlighted the significant risks associated with biased systems across various applications. Numerous studies have demonstrated that machine learning systems can exhibit biases related to demographic attributes such as gender, ethnicity, age, and geographical distribution, leading to unequal treatment of disadvantaged or underrepresented subpopulations. While fairness in machine learning has been extensively investigated in decision-making contexts such as job hiring, credit scoring, and criminal justice, it is only recently that researchers have begun to explore and address bias in medical image computing (MIC) and computer-assisted interventions (CAI).

To further this important discussion, the *2nd International Workshop on Fairness of AI in Medical Imaging (FAIMI 2024)* was held in Marrakesh, Morocco, on October 10, 2024. This workshop was organized in conjunction with the International Conference on Medical Image Computing and Computer Assisted Interventions (MICCAI 2024). The workshop aimed to raise awareness about potential fairness issues in machine learning within the context of biomedical image analysis. Additionally, it sought to bring together researchers from the MIC, CAI, machine learning, and fairness communities to discuss bias assessment and mitigation strategies.

The workshop featured three key sessions: (1) a keynote presentation by an expert speaker; (2) oral presentations by the authors of selected papers; and (3) poster presentations. All accepted papers were presented as posters, and attendees voted to determine the recipient of the Best Paper Award.

The peer-reviewed papers were selected using the CMT tool through a double-blind review process. Reviewers were selected through an open call for reviewers via the FAIMI newsletter, which led to a large and diverse pool of 37 reviewers from 32 different institutions on 5 different continents, ranging from first-time reviewers to those with more than 3 years of reviewing experience. To ensure fair assessment of all papers while simultaneously giving the diverse group of first-time reviewers a good learning experience, all papers were assigned five independent reviewers with a mix of experienced and new reviewers per paper, with careful consideration given to potential conflicts of interest and recent collaborations among peers. The reviews were overseen by the organizing committee at a paper selection meeting, where of the 17 papers submitted, 15 were accepted for publication, and the top five were invited to deliver oral presentations.

We would like to express our gratitude to our program committee members, authors, and attendees, whose contributions made FAIMI 2024 a resounding success. We would also like to thank the FAIMI Steering Committee for their advice and guidance and the

Activity Committee for their tireless work in delivering many of the practical aspects of workshop organisation.

October 2024

Esther Puyol-Antón
Aasa Feragen
Andrew P. King
Enzo Ferrante
Veronika Cheplygina
Melanie Ganz-Benjaminsen
Ben Glocker
Eike Petersen

FAIMI Organization

Program and Organizing Committee

Veronika Cheplygina — IT University Copenhagen, Denmark
Aasa Feragen — DTU Compute, Technical University of Denmark, Denmark
Enzo Ferrante — CONICET, Universidad Nacional del Litoral, Argentina
Melanie Ganz-Benjaminsen — University of Copenhagen, Neurobiology Research Unit, Rigshospitalet, Denmark
Ben Glocker — Imperial College London, UK
Andrew King — King's College London, UK
Eike Petersen — DTU Compute, Technical University of Denmark, Denmark
Esther Puyol-Antón — HeartFlow and King's College London, UK

Steering Committee

Judy Gichoya — Emory University, USA
Kanwal Bhatia — Aival, UK
Tal Arbel — McGill University, USA
Bishesh Khanal — NAAMII, Nepal
Ira Ktena — Deep Mind, UK
Sanmi Koyejo — Stanford University, USA
Karim Lekadir — Universitat de Barcelona, Spain

Activity Committee

Tareen Dawood — King's College London, UK
Nina Weng — DTU Compute, Technical University of Denmark, Denmark
Tiarna Lee — King's College London, UK
Dewinda Julianensi Rumala — Institut Teknologi Sepuluh Nopember, Indonesia
Emma Stanley — University of Calgary, Canada

Program Committee

Nina Weng	DTU Compute, Technical University of Denmark, Denmark
Annika Reinke	German Cancer Research Center, Germany
Cosmin Bercea	Technical University of Munich, Germany
Emma Stanley	University of Calgary, Canada
Tiarna Lee	King's College London, UK
Tareen Dawood	King's College London, UK
Zhen Yuan	King's College London, UK
Amelia Jiménez-Sánchez	University of Copenhagen, Denmark
Hilde Weerts	Eindhoven University of Technology, The Netherlands
Mahesan Niranjan	University of Southampton, UK
Po-Chih Kuo	National Tsing Hua University, Taiwan
Maria A. Zuluaga	EURECOM, France
Emese Sükei	Medical University of Vienna, Austria
Zeyan Liu	University of Kansas, USA
Soumya Kundu	King's College London, UK
Oscar Jimenez del Toro	Idiap Research Institute, Switzerland
Dewinda Julianensi Rumala	Institut Teknologi Sepuluh Nopember, Indonesia
Zikang Xu	University of Science and Technology of China, China
Didem Stark	Bernstein Center for Computational Neuroscience Berlin, Germany
Lemuel Puglisi	University of Catania, Italy
Estanislao Claucich	Universidad Nacional del Litoral, Argentina
Kate Cevora	Imperial College London, UK
Roshan Rane	Bernstein Center for Computational Neuroscience Berlin, Germany
Frank Li	Emory University, USA
Xiaohao Cai	University of Southampton, UK
Pratinav Seth	Arya.ai, India
Stella Grasshof	IT University Copenhagen, Denmark
Melat Ilmeya	Assam Science and Technology University, India
Gökhan Özbulak	École Polytechnique Fédérale de Lausanne, Switzerland

Azmine Toushik Wasi	University of Science and Technology, Bangladesh
Dimitri Kessler	University of Barcelona, Spain
Kaspar Ludvigsen	Durham University, UK
Fahad Shamshad	Mohamed bin Zayed University of Artificial Intelligence, United Arab Emirates

EPIMI Preface

The third *Ethical & Philosophical Issues in Medical Imaging* (EPIMI) workshop investigated questions that underlie medical imaging research at the most fundamental level. These investigations bridge the purely technical considerations traditionally seen in medical imaging research with humanistic ones and expand their scope beyond *what* is done towards *why* it is done. By doing so, we may more thoroughly understand our basic assumptions in performing medical imaging research and move towards dramatic answers to questions we never before thought to ask.

This instance of EPIMI concentrated on topics surrounding open science, turning a critical lens on the subject. Open science has been thought of as a mechanism for ensuring reproducible and honest research that is accessible to all. However, deeper investigation shows that mere openness is not a panacea for the scientific community. Souza et al. critique open research in artificial intelligence on the basis of global equity, providing a perspective that nuances how we can view fairness in medical machine learning research. Rosenblatt et al. critique the relationship between open datasets and trust in said datasets, showing how adversarial examples could be used to subtly change open datasets to disguise data fraud. These investigations show that even the concept of open science needs to be refined to best meet the needs of our community.

Papers were selected using a double-blind review process with author rebuttal through CMT. Each paper was evaluated by at least two independent reviewers in terms of the topicality of its content, the clarity of its presentation, the theoretical quality of its arguments, and the empirical quality of the evidence used to support them. Authors were then given an opportunity to critique their review and suggest changes to their final paper. Of the five papers submitted, two were selected for the EPIMI 2024 workshop, giving an acceptance rate of 40%. Authors of accepted papers were invited to give an oral presentation followed by an extended discussion period.

Lastly, we would to thank all those who submitted papers, the reviewers who critiqued them, and all those who contributed to the discussion at EPIMI 2024.

October 2023

John S. H. Baxter
Islem Rekik
Roy Eagleson
Ghada Zamzmi
Luping Zhou

EPIMI Organization

Program and Organizing Committee

John S. H. Baxter   Université de Rennes, France
Islem Rekik   Imperial College London, UK
Roy Eagleson   Western University, Canada
Ghada Zamzmi   Food and Drug Administration, USA
Luping Zhou   University of Sydney, Australia

Steering Committee

Elisabetta Lalumera   Università di Bologna, Italy
Pierre Jannin   Université de Rennes, France
Terry Peters   Western University, Canada
Dinggang Shen   Shanghai Technical University, China

Contents

EPIMI

FAIMI

Slicing Through Bias: Explaining Performance Gaps in Medical Image Analysis Using Slice Discovery Methods

Vincent Olesen[1], Nina Weng[1], Aasa Feragen[1], and Eike Petersen[1,2]

[1] Technical University of Denmark, Kongens Lyngby, Denmark
{ninwe,afhar,ewipe}@dtu.dk
[2] Fraunhofer Institute for Digital Medicine MEVIS, Bremen, Germany

Abstract. Machine learning models have achieved high overall accuracy in medical image analysis. However, performance disparities on specific patient groups pose challenges to their clinical utility, safety, and fairness. This can affect known patient groups – such as those based on sex, age, or disease subtype – as well as previously unknown and unlabeled groups. Furthermore, the root cause of such observed performance disparities is often challenging to uncover, hindering mitigation efforts. In this paper, to address these issues, we leverage Slice Discovery Methods (SDMs) to identify interpretable underperforming subsets of data and formulate hypotheses regarding the cause of observed performance disparities. We introduce a novel SDM and apply it in a case study on the classification of pneumothorax and atelectasis from chest x-rays. Our study demonstrates the effectiveness of SDMs in hypothesis formulation and yields an explanation of previously observed but unexplained performance disparities between male and female patients in widely used chest X-ray datasets and models. Our findings indicate shortcut learning in both classification tasks, through the presence of chest drains and ECG wires, respectively. Sex-based differences in the prevalence of these shortcut features appear to cause the observed classification performance gap, representing a previously underappreciated interaction between shortcut learning and model fairness analyses.

Keywords: Slice Discovery Methods · Algorithmic Fairness · Shortcut Learning · Chest X-ray · Model Debugging

1 Introduction

Machine learning models have shown great promise in medical image-based diagnosis, sometimes with performance claims that rival human experts. However, reported performance may overstate these models' clinical utility and safety [29].

Supplementary Information The online version contains supplementary material available at https://doi.org/10.1007/978-3-031-72787-0_1.

Specifically, models may underperform or fail systematically on critical subsets of data even while overall average accuracy remains high. In computer vision research, such subsets are called *underperforming slices* or *blind spots* [7,22], where 'slice' refers to a subset of samples with similar characteristics, such as an attribute familiar to a domain expert.[1]

In medical image analysis, a model might be underperforming on a slice of patients for a wide range of reasons, including group under-representation, increased input or label noise, fundamental differences in the difficulty of the prediction task, and shortcut learning [5,11,19,21]. Performance disparities between patient groups have been observed in many medical imaging domains [4,8,16,24,32], raising concerns about the potential unfairness resulting from the application of such models. However, properly *mitigating* such performance disparities requires identifying their root cause, which is often challenging [18,21,28]. The challenge is further compounded by the fact that the feature that causally distinguishes high-performing from low-performing patients is often unknown and, thus, not annotated. This renders simple subgroup analyses based on available metadata insufficient for identifying the causes of performance disparities.

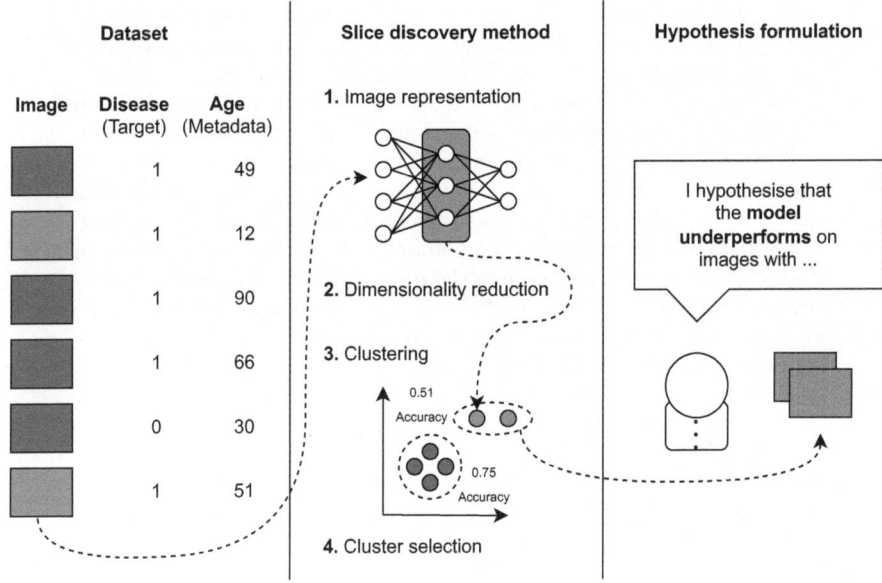

Fig. 1. A general overview of the key elements of slice discovery methods.

---

[1] In the medical imaging literature, the term 'slice' commonly refers to a two-dimensional cross-section within three-dimensional volumetric data. We are adopting a differing terminology from earlier work on SDMs originating outside of the medical image analysis field. We apologize for the unfortunate clash of terminology.

To address the issue of unknown distinguishing features, various methods for the unsupervised discovery of underperforming slices have been proposed in the computer vision literature. Such methods are variously known as *Slice Discovery Methods* (SDMs) [7] or *Blindspot Discovery Methods* (BDMs) [22]. Typically, these methods perform a cluster analysis on the input space and then select poorly performing clusters, or *slices* of data, for further analysis; refer to Fig. 1 for a high-level overview. SDMs can aid machine learning practitioners and domain experts in identifying underperforming sets of data, as well as in forming hypotheses about the *causes* of this underperformance. With few exceptions [19], SDMs have not yet found widespread use in the medical imaging domain.

In this study, we explore the use of SDMs for the analysis of performance disparities in medical image analysis. Our contributions are twofold. First, we provide a general overview of SDMs in medical image analysis and we propose a novel SDM, rigorously justifying all of our design choices. We demonstrate the effectiveness of our proposed SDM for hypothesis formulation in a case study of pneumothorax and atelectasis classification on two public chest X-ray datasets (NIH-CXR 14 [27] and CheXpert [9]). Second, by further analyzing the hypotheses generated using our SDM, we show that chest drain shortcut learning causes a previously unexplained yet variously reproduced performance gap between male and female subjects in pneumothorax classification. This constitutes an important link between shortcut learning and model fairness analyses that has, to the authors' knowledge, not been described before. In addition, using our SDM, we discover a new shortcut feature (the presence of ECG cables) that may explain male–female performance disparities in atelectasis classification.

2 Related Work

2.1 Bias and Shortcuts in Chest X-Ray Analysis

Algorithmic fairness in medical image analysis, and performance disparities between patient groups in particular, have recently come under rapidly increasing scrutiny [4,14,19,23,24,28]. In this context, the fairness of chest x-ray-based disease classification models has received particularly broad attention [2,8,14,24,28,30]. Larrazabal et al. [14] demonstrated that such models had better classification performance for a particular patient group (based on biological sex) if that group was represented in higher proportions of the training dataset. While not the focus of their study, their results also indicated significant differences between model performance on male and female subjects, with the classification models performing better for either group in different diseases. These (sometimes large) performance gaps persisted even in the case of sex-balanced training sets. This observation prompted Weng et al. [28] to investigate the hypothesis that biological sex differences were causing these unexplained performance gaps. Based on their results, the authors dismissed breast shadows as a factor, but other biological sex differences contributing to performance gaps remain uncertain, leading to an unexplained gender disparity. Zhang et al. [30] employed standard algorithmic fairness mitigation approaches to the chest x-ray case, finding that simple group balancing was one of the most robust approaches – which did not, however, mitigate the performance gaps observed by Larrazabal et al. [14].

In a separate development, it has been widely demonstrated that chest x-ray-based disease classification models are prone to relying on shortcut learning [2,5,11,19]. Both Oakden-Rayner et al. [19] and Jiménez-Sánchez et al. [11] demonstrate how pneumothorax classification tends to rely heavily on the presence of chest drains, which represent the standard treatment for pneumothorax. Connecting the two challenges of shortcut learning and fairness, many authors have raised concerns about the potential for deep learning models to exploit spurious correlations between sensitive attributes, such as age, gender, or ethnicity, and the prediction target [2,8]. To the authors' knowledge, the fact that shortcut learning relying on *non*-sensitive features (such as the presence of chest drains) can explain performance disparities between sensitive groups (such as gender groups) has not been discussed explicitly before. Notably, Jiménez-Sánchez et al. [11] took important first steps in this direction, by differentially reporting the effect of shortcut learning on different gender groups.

2.2   Slice Discovery Methods

Slice discovery methods (SDMs) are a recently emerging tool for the performance analysis and subsequent improvement of deep learning models. In particular, they aim to solve the problem that the features that identify underperforming groups of inputs might not be known a priori. To this end, SDMs typically perform unsupervised clustering of the input data, in order to identify semantically similar 'slices' of data that the model under analysis performs poorly on. In more detail, SDMs usually consist of the following steps: (1) the input data is embedded into a latent representation space, (2) some SDMs perform dimensionality reduction, (3) an unsupervised learning method, such as clustering, is used to extract slices of the data, and (4) the extracted slices are prioritized based on a performance metric, such as accuracy. In Table 1, adapted from Plumb et al. [22], we summarize previously proposed SDMs and their respective high-level design.

The most common methods to extract slices are clustering algorithms, namely the Gaussian Mixture Model (GMM) and its variants. The most common choice of image embedding is the latent space representation computed by the classification model under scrutiny, i.e., its penultimate layer's outputs. However, recent methods instead utilize multi-modal pre-trained models like CLIP to enable the generation of text descriptions for extracted slices [7,10]. Interestingly, the crucial dimensionality reduction step has received relatively little attention yet, as recently pointed out by Plumb et al. [22]. Possibly due to their relatively recent emergence, SDMs have not yet been widely applied in the medical image domain. In this regard, the work of Oakden-Rayner et al. [19] represents a very notable early exception that precedes more recent SDM developments.

Table 1. A summary of slice discovery methods. Clf: The representation used by the classification model under analysis. Adapted from Plumb et al. [22].

Method	Rep.	Dim. reduction	Clustering
Algorithmic measurement [19]	Clf		KNN
MultiAccuracy Boost [13]	VAE		Rigid/decision-tree regression
GEORGE [25]	Clf/BiT emb.	UMAP $(d=1,2)$	GMM
Spotlight [6]	Clf		Optimization problem
Planespot [22]	Clf	scvis $(d=2)$	GMM
Domino [7]	CLIP	PCA $(d=128)$	GMM
Failure mode distillation [10]	CLIP		SVM
Bias-Aware Hierarchical Clustering [17]	Clf	UMAP$(d=2)$	Mod. K-Means
Proposed SDM (Ours)	Clf	FC layer $(d=128)$	GMM

3 Methodology

3.1 Proposed Slice Discovery Method

This section introduces our proposed SDM and motivates our design choices. The proposed method consists of the following four steps, following Fig. 1:

Image representation. We use the image representation computed by the penultimate layer of the classification model under scrutiny. As opposed to SDMs that use a separate model to obtain image embeddings, our approach relies only on information available to the model's final classification layer.

Dimensionality reduction. We insert a single fully connected layer with d-dimensional output (and a sigmoid activation layer) between the classification model's penultimate layer and its final classification head. We train (just) this additional layer following the same procedure that was used for the classification model itself. Similarly to the previous step and contrary to standard choices such as PCA, t-SNE, or UMAP, our approach is *supervised* and preferably preserves information relevant to the model's predictions.

Clustering. We use a Gaussian Mixture Model (GMM) for clustering and the Bayesian Information Criterion (BIC) for choosing the number of clusters. We cluster disease-positive and -negative samples separately to extract clusters of the same error type, similar to Oakden-Rayner et al. [19].

Cluster selection. We propose using the Brier score (BS), a proper scoring rule, to prioritize under- and overperforming slices, equivalent to mean squared error between model confidence and binary target labels. The main motivation for our choice is that, as opposed to classification accuracy, the BS is threshold-independent. In addition, it captures both the model's discriminative ability and its calibration [1]. As opposed to AUROC [12], per-cluster

BS can be meaningfully compared between groups, and as opposed to many calibration metrics [20,23], it can be meaningfully compared between clusters of different sizes. We quantify BS uncertainty by simple bootstrapping of each cluster and select the cluster with the lowest 97.5-quantile as the best, and the cluster with the highest 2.5-quantile as the worst.

3.2 Datasets

We consider a case study on two public datasets, NIH-CXR14 [27] and CheX-pert [9]. Both datasets slightly over-represent male subjects. We reused chest drain labels previously crowd-sourced from both radiologists and nonexperts [3,11,19], including 3543 random cases with pneumothorax in the NIH dataset and 972 cases with and without pneumothorax in CheXpert. For NIH, we observe a larger prevalence of chest drains among pneumothorax-positive male subjects compared to female subjects (49.5% vs. 42.8%). For CheXpert, we observe a larger prevalence of chest drains among pneumothorax-negative male subjects (23.0% vs. 14.7% in females) but comparable chest drain prevalence across sexes for pneumothorax-positive subjects (50.5% in males vs. 50.2% in females).

3.3 Experiments

We conduct a case study on the pneumothorax classification task, following the experimental setup of Weng et al. [28]. Specifically, to reduce the potential for label noise to affect our analyses, we select one sample per patient, with a preference for pneumothorax-positive samples and an equal sex ratio. Withholding the chest drain-annotated samples, we split the datasets into 60%/10%/30% train, validation, and test sets, resampling the splits ten times. We resize the images to 224×224 pixels and train a ResNet50 (Adam optimizer, learning rate 10^{-6}, batch size 64, 20 epochs). We use data augmentation for the training dataset, including horizontal flipping, rotation up to 15°, and scaling up to 10% with a 50% probability for each augmentation. We then carry out our proposed SDM and report the distribution of comorbidities and chest drains (for annotated chest drain samples).[2] We repeat the same analyses for atelectasis classification. Based on our findings, we conduct further post-hoc analyses in both cases. Our source code is publicly available at https://github.com/volesen/slicing-through-bias.

4   Results

Pneumothorax Classification. Figure 2 and Fig. A.1 in the supplementary material show the best- and worst-performing slices based on the Bier score in the

---

[2] For the dimensionality reduction, we used d = 128 following similar previous work [7]. Results with d = 10 were comparable. The gap statistic [26] indicates that clustering does indeed occur in the reduced space. The effect of the additional dimensionality reduction layer on the model's classification performance was negligible in terms of test accuracy and AUROC.

NIH-CXR14 and CheXpert datasets. In both datasets, the underperforming slices for pneumothorax-negative samples have a lower-than-average chest drain proportion, while the opposite holds for pneumothorax-positive cases.

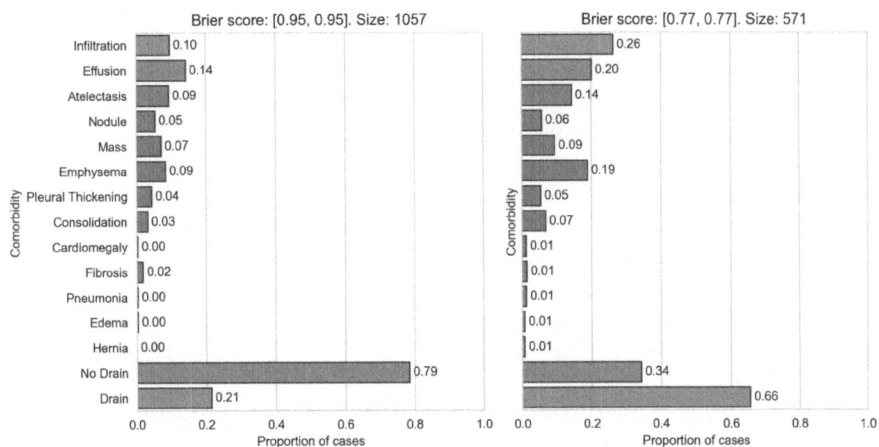

Fig. 2. The comorbidity and chest drain distribution in pneumothorax-positive chest drain annotated samples of NIH-CXR14 for the worst-performing (left column) and best-performing (right column) slices by Brier score. The pneumothorax-negative case is omitted as chest drain annotations were not available for these samples in NIH-CXR14.

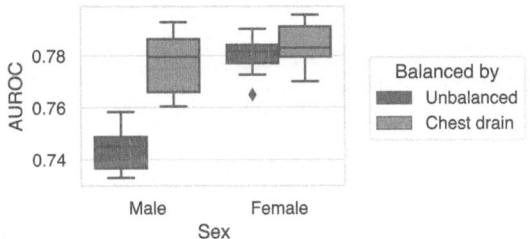

Fig. 3. AUROC on CheXpert with male and female test subjects on pneumothorax prediction, following the natural ('unbalanced') distribution of chest drains and balanced by chest drain presence across ten samplings of the train-validation-test sets.

Based on these results (and based on prior work [11,19]), we hypothesized that the presence of chest drains is used to classify pneumothorax. Indeed, we observe that computed model confidence (the softmax output of the model for the disease-positive class) is consistently higher in samples with chest drains across datasets, sexes, and pneumothorax labels (positive/negative), lending strong support to this hypothesis (Fig. A.2 in the supplementary material). Furthermore, as both datasets have varying chest drain prevalences by sex, we hypothesized that this could contribute to the gender performance gap. To assess this hypothesis, we determined AUROC by sex, firstly in a test set following the natural distribution of chest drains, and secondly a test set with equalized chest drain prevalence in the male and female populations (Fig. 3). We observe a significant difference in performance in the first case ($p < 0.001$) but no significant difference in the second, chest-drain-balanced case ($p > 0.1$), indicating that chest drain shortcut learning is the cause of a large part of the male–female performance gap observed in prior work [14,28]. Statistical significance was assessed using a Mann-Whitney U-test. It should be noted that the differences in comorbidities noticeable in Fig. 2 represent another potential explanation for the observed performance disparities. While we did not further explore this hypothesis here, our additional experiments related to the chest drain shortcut hypothesis (discussed above) seem to indicate that this indeed explains the bulk of the observed performance gaps.

Atelectasis Classification. We repeated our SDM analysis for atelectasis. (Slice statistics not shown here due to space constraints.) Prompted by a visual inspection of the recordings in the best and worst-performing slices, we hypothesized that the two slices differed substantially in their prevalence of ECG cables in the recordings. To test this hypothesis, we randomly selected and (as non-experts) labeled 100 samples each for the best- and worst-performing slices on atelectasis-positive and -negative cases according to whether they displayed ECG cables or not. Of atelectasis-negative samples, 95% of the labeled recordings in the worst-performing slice contained ECG cables, compared to 10% in the best-performing slice. Of the atelectasis-positive cases, 50% of the labeled recordings in the worst-performing slice had ECG cables, compared to 99% in the best-performing slice.

5   Discussion and Conclusion

We have proposed a novel slice discovery method (SDM), which differs from previously proposed methods in several key elements. Both in the representation extraction as well as in the (supervised) dimensionality reduction step, we prioritize only using information available to the classification model under test. In addition, we propose using the Brier score (BS) for selecting highly and poorly performing clusters because it is threshold-independent, accounts for both discriminative ability and calibration, and enables meaningful comparisons between clusters of different sizes. In a case study on chest x-ray-based disease classification, our SDM successfully recovered a previously known case of shortcut learning (chest drains for pneumothorax classification) and suggested a new, previously unknown case (ECG cables for atelectasis classification). The latter case also demonstrated another benefit of using SDMs: reducing the required labeling efforts, because these can be specifically targeted at the worst and best clusters. Our case study shows that our proposed SDM, and SDMs in general, can aid researchers in generating hypotheses regarding the causes of model underperformance on subsets of data, which is crucial for leveling *up* performance [21].

Consistent with the observations made by Weng et al. [28], our findings challenge the notion that biological differences are the primary driver of the previously observed [14] but unexplained male–female performance gaps in chest x-ray-based disease classification. Instead, our results suggest that shortcut learning in conjunction with a difference in chest drain prevalence between males and females causes the observed performance disparity. This newly gained knowledge opens up the possibility for the targeted application of shortcut learning mitigation techniques, instead of relying on blind group performance equalization approaches that often result in leveling *down* performance [21,30,31].

While our study builds upon the chest drain annotations of [3] and [11], for which the former show a high level of agreement between expert and non-expert annotations, caution is warranted. Non-expert annotations, although showing agreement, may not necessarily represent the ground truth or offer a representative sample (images without consensus between multiple labelers were disregarded [11]). This caution is emphasized by a significant difference in the prevalence of chest drains among pneumothorax-positive samples in NIH-CXR14 between studies [3,19]. Moreover, the mitigation of shortcut learning is a highly active research area, and mitigating many different shortcuts simultaneously, and with limited label availability, remains challenging [15]. Finally, the interpretation of identified slices in more challenging cases represents a crucial challenge for future research. Both chest drains and ECG cables are visible to the human eye (though maybe not the non-expert), but other potentially problematic features may not be.

Acknowledgements. Work on this project was partially funded by the Independent Research Fund Denmark (DFF, grant number 9131-00097B), Denmark's Pioneer Centre for AI (DNRF grant number P1), and the Novo Nordisk Foundation through the Center for Basic Machine Learning Research in Life Science (MLLS, grant number NNF20OC0062606). The funding agencies had no influence on the writing of the manuscript nor on the decision to submit it for publication.

References

1. Bröcker, J.: Reliability, sufficiency, and the decomposition of proper scores. Q. J. R. Meteorol. Soc. **135**(643), 1512–1519 (2009)
2. Brown, A., Tomasev, N., Freyberg, J., Liu, Y., Karthikesalingam, A., Schrouff, J.: Detecting shortcut learning for fair medical AI using shortcut testing. Nat. Commun. **14**(1) (2023)
3. Damgaard, C., Eriksen, T.N., Juodelyte, D., Cheplygina, V., Jiménez-Sánchez, A.: Augmenting chest x-ray datasets with non-expert annotations (2023)
4. Daneshjou, R., Vodrahalli, K., Novoa, R.A., Jenkins, M., Liang, W., Rotemberg, V., et al.: Disparities in dermatology AI performance on a diverse, curated clinical image set. Sci. Adv. **8**(32) (2022)
5. DeGrave, A.J., Janizek, J.D., Lee, S.I.: AI for radiographic COVID-19 detection selects shortcuts over signal. Nat. Mach. Intell. **3**(7), 610–619 (2021)

6. d'Eon, G., d'Eon, J., Wright, J.R., Leyton-Brown, K.: The spotlight: a general method for discovering systematic errors in deep learning models. In: ACM FAccT, pp. 1962–1981 (2022)

7. Eyuboglu, S., Varma, M., Saab, K.K., Delbrouck, J.B., Lee-Messer, C., Dunnmon, J., et al.: Domino: discovering systematic errors with cross-modal embeddings. In: ICLR (2022)

8. Glocker, B., Jones, C., Roschewitz, M., Winzeck, S.: Risk of bias in chest radiography deep learning foundation models. Radiol. Artif. Intell. **5**(6) (2023)

9. Irvin, J., Rajpurkar, P., Ko, M., Yu, Y., Ciurea-Ilcus, S., Chute, C., et al.: CheXpert: a large chest radiograph dataset with uncertainty labels and expert comparison. In: AAAI/IAAI/EAAI (2019)

10. Jain, S., Lawrence, H., Moitra, A., Madry, A.: Distilling model failures as directions in latent space. In: ICLR (2023)

11. Jiménez-Sánchez, A., Juodelyte, D., Chamberlain, B., Cheplygina, V.: Detecting shortcuts in medical images - a case study in chest X-rays. In: ISBI, Cartagena, Colombia. IEEE (2023)

12. Kallus, N., Zhou, A.: The fairness of risk scores beyond classification: bipartite ranking and the xAUC metric. In: NeurIPS, vol. 32. Curran Associates, Inc. (2019)

13. Kim, M.P., Ghorbani, A., Zou, J.: Multiaccuracy: black-box post-processing for fairness in classification. In: AIES, pp. 247–254. ACM (2019)

14. Larrazabal, A.J., Nieto, N., Peterson, V., Milone, D.H., Ferrante, E.: Gender imbalance in medical imaging datasets produces biased classifiers for computer-aided diagnosis. Proc. Natl. Acad. Sci. **117**(23), 12592–12594 (2020)

15. Li, Z., Evtimov, I., Gordo, A., Hazirbas, C., Hassner, T., Ferrer, C.C., et al.: A whac-a-mole dilemma: shortcuts come in multiples where mitigating one amplifies others. In: CVPR, pp. 20071–20082 (2023)

16. Lin, M., Li, T., Yang, Y., Holste, G., Ding, Y., Van Tassel, S.H., et al.: Improving model fairness in image-based computer-aided diagnosis. Nat. Commun. **14**(1) (2023)

17. Misztal-Radecka, J., Indurkhya, B.: Bias-aware hierarchical clustering for detecting the discriminated groups of users in recommendation systems. Inf. Process. Manag. **58**(3), 102519 (2021)

18. Mukherjee, P., Shen, T.C., Liu, J., Mathai, T., Shafaat, O., Summers, R.M.: Confounding factors need to be accounted for in assessing bias by machine learning algorithms. Nat. Med. **28**(6), 1159–1160 (2022)

19. Oakden-Rayner, L., Dunnmon, J., Carneiro, G., Re, C.: Hidden stratification causes clinically meaningful failures in machine learning for medical imaging. In: CHIL, pp. 151–159. ACM (2020)

20. Petersen, E., Ganz, M., Holm, S.H., Feragen, A.: On (assessing) the fairness of risk score models. In: FAccT. ACM (2023)

21. Petersen, E., Holm, S., Ganz, M., Feragen, A.: The path toward equal performance in medical machine learning. Patterns **4**(7) (2023)

22. Plumb, G., Johnson, N., Cabrera, A., Talwalkar, A.: Towards a more rigorous science of blindspot discovery in image classification models. Trans. Mach. Learn. Res. (2023)

23. Ricci Lara, M.A., Mosquera, C., Ferrante, E., Echeveste, R.: Towards unraveling calibration biases in medical image analysis. In: Wesarg, S., et al. (eds.) CLIP EPIMI FAIMI 2023. LNCS, vol. 14242, pp. 132–141. Springer, Cham (2023). https://doi.org/10.1007/978-3-031-45249-9_13

24. Seyyed-Kalantari, L., Zhang, H., McDermott, M.B.A., Chen, I.Y., Ghassemi, M.: Underdiagnosis bias of artificial intelligence algorithms applied to chest radiographs in under-served patient populations. Nat. Med. **27**(12), 2176–2182 (2021)

25. Sohoni, N., Dunnmon, J., Angus, G., Gu, A., Ré, C.: No subclass left behind: fine-grained robustness in coarse-grained classification problems. In: NeurIPS, vol. 33, pp. 19339–19352 (2020)

26. Tibshirani, R., Walther, G., Hastie, T.: Estimating the number of clusters in a data set via the gap statistic. J. R. Stat. Soc. Ser. B Stat. Methodol. **63**(2), 411–423 (2001). https://doi.org/10.1111/1467-9868.00293

27. Wang, X., Peng, Y., Lu, L., Lu, Z., Bagheri, M., Summers, R.M.: ChestX-Ray8: hospital-scale chest x-ray database and benchmarks on weakly-supervised classification and localization of common thorax diseases. In: CVPR, pp. 2097–2106 (2017)

28. Weng, N., Bigdeli, S., Petersen, E., Feragen, A.: Are sex-based physiological differences the cause of gender bias for chest x-ray diagnosis? In: Wesarg, S., et al. (eds.) CLIP EPIMI FAIMI 2023. LNCS, vol. 14242, pp. 142–152. Springer, Cham (2023). https://doi.org/10.1007/978-3-031-45249-9_14

29. Wynants, L., Van Calster, B., Collins, G.S., Riley, R.D., Heinze, G., Schuit, E., et al.: Prediction models for diagnosis and prognosis of covid-19: systematic review and critical appraisal. BMJ m1328 (2020)

30. Zhang, H., Dullerud, N., Roth, K., Oakden-Rayner, L., Pfohl, S., Ghassemi, M.: Improving the fairness of chest x-ray classifiers. In: Conference on Health, Inference, and Learning, pp. 204–233. PMLR (2022)

31. Zietlow, D., et al.: Leveling down in computer vision: pareto inefficiencies in fair deep classifiers. In: CVPR. IEEE (2022)

32. Zong, Y., Yang, Y., Hospedales, T.: MEDFAIR: benchmarking fairness for medical imaging. In: ICLR (2023)

Dataset Distribution Impacts Model Fairness: Single Vs. Multi-task Learning

Ralf Raumanns[1,2]([✉]), Gerard Schouten[1], Josien P. W. Pluim[2],
and Veronika Cheplygina[3]

[1] Fontys University of Applied Science, Eindhoven, The Netherlands
ralf.raumanns@gmail.com
[2] Eindhoven University of Technology, Eindhoven, The Netherlands
[3] IT University of Copenhagen, Copenhagen, Denmark

Abstract. The influence of bias in datasets on the fairness of model predictions is a topic of ongoing research in various fields. We evaluate the performance of skin lesion classification using ResNet-based CNNs, focusing on patient sex variations in training data and three different learning strategies. We present a linear programming method for generating datasets with varying patient sex and class labels, taking into account the correlations between these variables. We evaluated the model performance using three different learning strategies: a single-task model, a reinforcing multi-task model, and an adversarial learning scheme. Our observations include: 1) sex-specific training data yields better results, 2) single-task models exhibit sex bias, 3) the reinforcement approach does not remove sex bias, 4) the adversarial model eliminates sex bias in cases involving only female patients, and 5) datasets that include male patients enhance model performance for the male subgroup, even when female patients are the majority. To generalise these findings, in future research, we will examine more demographic attributes, like age, and other possibly confounding factors, such as skin colour and artefacts in the skin lesions. We make all data and models available on GitHub.

Keywords: Skin lesions · Bias · Fairness · Multi-task learning · Adversarial learning

1 Introduction

Deep learning has shown many successes in medical image diagnosis [3,12,30], but despite high overall performance, models can be biased against patients from different demographic groups [1,15,22]. Bias and fairness are becoming an active topic in medical imaging, with studies focusing for instance on skin lesions [1,17], chest x-rays [22] and brain MR scans [27]. Examples of sensitive attributes

Supplementary Information The online version contains supplementary material available at https://doi.org/10.1007/978-3-031-72787-0_2.

include age, sex or race. For skin lesion classification, the Fitzpatrick skin type is often studied [4,17,31,34].

Fairness studies typically include baselines showing bias between groups, and/or propose methods to improve fairness. The methods are based on sampling or weighting strategies during training [17], and/or introducing training strategies that try to debias the methods to rely on the sensitive attributes, such as adversarial methods [1]. For instance, Yang and colleagues [35] developed an adversarial debiasing framework to reduce biases in hospital location and patient ethnicity. Similarly, Wu et al. introduced FairPrune [34], a method for pruning parameters based on their significance to both privileged and unprivileged groups, focusing on sex and skin tone. Moreover, Bevan and Atapour-Abarghouei [5] employed various strategies to limit bias in skin lesion images, specifically targeting discrepancies arising from medical instruments, surgical markings, and rulers. Popular datasets include ISIC skin lesion datasets [9–11,18,29,32] and Fitzpatrick-17K [16,17]. Researchers either use already provided data splits for evaluation, or split the data ratios of patients with a specific demographic attribute, for example male vs female patients.

Our current study builds on two crucial insights from other topics in medical imaging: multi-task learning and shortcut learning [13,25]. Firstly, some studies use demographic attributes within multi-task learning settings; for example, [23]. Here the attributes are *reinforcing* the diagnosis during optimization. This is at odds with the more recent adversarial strategies [1,2] where models are encouraged to NOT predict the sensitive attribute. Secondly, there are correlations between demographics and demographic attributes and shortcut learning, including, for example, imaging devices and surgical markers [5,6,15,21,33]. In such cases, simply splitting the data according to a specific attribute can create imbalance in terms of the other attributes, thus the observed (un)fairness could be due to the attributes that were not considered.

Our contributions are as follows:

1. We propose using the linear programming (LP) approach for skin lesions retrieved from the ISIC archive [9–11,18,29,32] via the gallery browser [20]. This method gives more control over patient subset assignment. It adjusts the proportions of selected dataset attributes while keeping others constant.
2. We systematically study two strategies that handle the demographic variable in different ways: a reinforcing multi-task strategy [24,28] and an adversarial strategy [1,2,8]
3. We evaluate our models using overall and subgroup Area Under the Curve (AUC) based on sex, and show that:
   - Models perform better for male subgroups in the male-only and lightly skewed male patient experiments. In the balanced dataset and lightly skewed female patient experiments, there is no significant difference between the subgroups. However, in the lightly skewed female patient scenario, the adversarial model performs better for male patients.
   - Models trained exclusively on female patients exhibit a positive difference in performance for female patients.

- The base model reveals a significant sex bias, performing worse for female patients, except when trained exclusively on female patients.
- The reinforcement model has no significant effect on sex bias.
- The adversarial model significantly reduces sex bias in scenarios involving only female patients.
- Eliminating model bias is challenging, with significant performance gaps observed in datasets with skewed sex distributions.
4. We make all data and models available on https://github.com/raumannsr/data-fairness-impact.

2   Methods

Construction of Datasets. We trained and validated our models on datasets with different female (F) to male (M) patient ratios, equal numbers of malignant and benign lesions, and equal number of patients below and above 60 (median age) for each sex. We refer to the datasets as M100 (100% male patients), F25M75 (25% female patients, 75% male patients), F50M50, F75M25, and F100 are defined analogously. We evaluate the models using a balanced test set mirroring F50M50.

We used the ISIC archive's [9–11, 18, 29, 32] gallery browser [20], which had 81,155 dermoscopic images of skin lesions, some with age and sex metadata. We queried the archive for "dermoscopic" images diagnosed as "benign" or "malignant" for all ages and both sexes. This gave us 71,035 images (62,439 benign, 8,596 malignant), which we processed using the steps in Fig. 1 (see Appendix for more details).

Linear Programming for Optimal Dataset Construction. We have developed a method to create diverse dataset compositions using linear programming, a common mathematical optimisation technique. The goal is to maximise the number of instances of skin lesions within defined constraints, as we express below:

Find a vector	x	(decision variables)
that maximises	$f = x_1$	(objective function)
subject to	$a_{i1}x_1 + a_{i2}x_2 + \cdots + a_{in}x_n \leq b_i$	(constraints)
	for $i = 1, \ldots, 13$	
and	$x_j \geq 0.$	(non-negativity constraints)
	for $j = 1, \ldots, 14$	

- The decision variables $(x_1, \ldots, x_{14})$ denote specific categories, like benign lesions in female patients aged 60 and above. See Appendix for more.
- The objective function of the LP model is designed to maximise the number of malignant instances x_1. There are fewer malignant instances than benign ones in the ISIC archive, and the goal is to achieve a balance between the two.

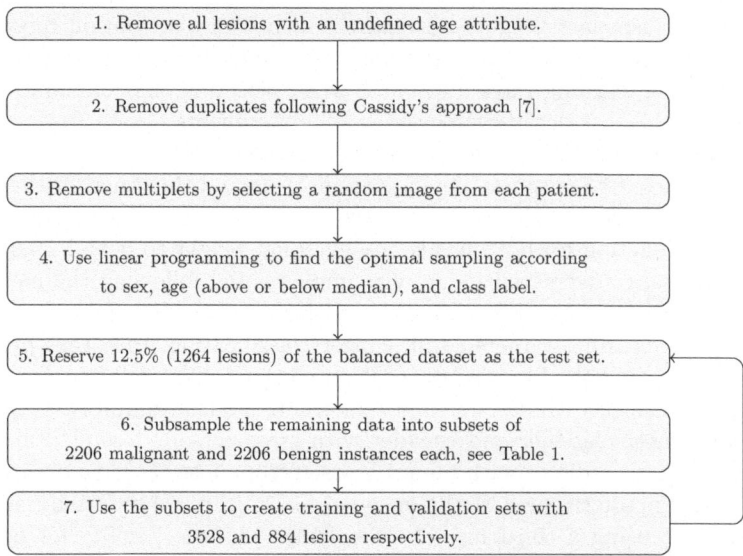

Fig. 1. Steps for filtering lesions and creating test, training and validation sets. Steps 5 through 7 are repeated using 5 different seeds in a cross-validation setup.

- The constraints limit the solution by setting specific limits for each group and maintaining ratios between these groups. Group examples include all benign lesions, all female patients over 60, and all male patients under 60. The primary constraint ensures an equal number of malignant and benign lesions $(x_1 - x_2 = 0)$. See Appendix for more.
- Due to non-negativity constraints, decision variables cannot be negative.

Table 1. Datasets are distributed amongst malignant, benign, male patients (M), and female patients (F) categories for both training and validation.

	M100	F25M75	F50M50	F75M25	F100
Malignant (M/F)	2206 (2206/0)	2941 (2206/735)	4412 (2206/2206)	3235 (809/2426)	2426 (0/2426)
Benign (M/F)	2206 (2206/0)	2941 (2206/735)	4412 (2206/2206)	3235 (809/2426)	2426 (0/2426)

Within set constraints, the optimal solution maximises malignant lesions and assigns value to decision variables. To find this solution, we created a unique LP model for each dataset. Table 1 shows the result of the LP model for the different datasets.

Models. We used the ResNet50 model [19] in three ways, which include:

- The single-task baseline model enhanced with two fully connected layers has a sigmoid activation function and binary cross-entropy loss (L_c, see Eq. 1).

We did not use class weights but rather the actual distribution represented by the training dataset. We fine-tuned the model through a grid search of three varying learning rates and batch sizes, selecting the combination that yielded the highest performance across all experiments.

– The multi-task reinforcing model, with three added layers to the convolutional base, produces two outputs: one for classification and the other for the binary sex attribute. We employed binary cross-entropy loss (L_c) and a sigmoid activation function for both heads, giving equal weight to both losses.

– The multi-task adversarial model was implemented following the methodology [1,2], using a network with a shared feature encoder and two classifier heads. One classifier targeted skin cancer classification; the other predicted confounding variables like sex or age. We used the ResNet architecture to compare performance with baseline and reinforcing models equitably. We trained the skin cancer classifier and encoder with cross-entropy loss (L_c) and optimised the bias predictor with binary cross-entropy loss (L_c). To diminish the confounder predictiveness of the encoded feature, we adversarially adjusted the encoder using a third loss (L_{br}), setting λ as the penalty for accurate demographic predictions. We used a lambda (λ, see Eq. 2) value of 5, as in [1], to assess subgroup performance and set the penalty for correct predictions of the target demographic parameter.

To summarise, we use these loss functions:

$$L_c = -\sum_{i=1}^{n}[y_i \log(\hat{y}_i) + (1 - y_i) \log(1 - \hat{y}_i)] \tag{1}$$

$$L_{br} = \lambda L_c \tag{2}$$

where n represents the number of lesions, and y_i and $\hat{y}_i$ denote respectively the prediction and the expected outcome for lesion i.

We pre-trained all networks with ImageNet and resized images to 384×384 for ResNet50's input size. During model training, we used data augmentation techniques, ran up to 40 epochs with a batch size of 20, and set a learning rate of $2.0e-5$. We stopped training if no significant improvement occurred after 10 consecutive epochs to avoid overfitting. We implemented our baseline and reinforcing models in Keras with the TensorFlow backend [14] and our adversarial model in PyTorch [26].

Evaluation. For the purpose of an in-depth evaluation, we generated five distinct instances of each dataset: M100, F25M75, F50M50, F75M25 and F100. Each instance was created with a unique seed to ensure diversity and robustness in our evaluation. Each seed corresponds to a balanced test set to allow fair comparisons between dataset-model instances. Furthermore by using a balanced test set, we ensure that the results are not skewed towards any specific subgroup. We evaluate AUC overall and for male and female subgroups for each model architecture and dataset combination.

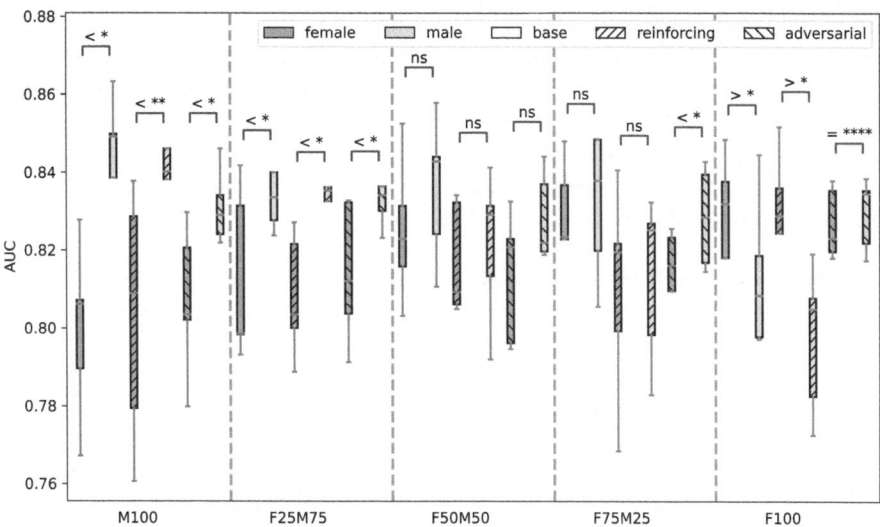

Fig. 2. The AUC score varies based on data splits ranging from only male patients (M100) to only female patients (F100). We show base, reinforcing and adversarial model performance for female and male patient subgroups. Significance per Mann-Whitney U test (as used in [22]) is denoted by **** ($P \leq 0.0001$), *** ($0.0001 < P \leq 0.001$), ** ($0.001 < P \leq 0.01$), * ($0.01 < P \leq 0.1$), and not significant (ns) ($P > 0.1$). $<$ indicates lower AUCs, $>$ higher AUCs, and $=$ comparable AUCs for female patients.

3   Results

Figure 2 shows the impact of dataset distributions on three learning strategies, reporting AUC scores overall and for both sexes.

Sex-Specific Training Data Yields Better Results. Models perform better for male subgroups in the male-only and lightly skewed male patient experiments. In the balanced dataset and lightly skewed female patient experiments, there is no significant difference between the subgroups. However, in the lightly skewed female patient scenario, the adversarial model performs better for male patients. An exception is observed when the training datasets consist only of female patients; in such cases, there is a positive difference in AUC scores favouring female patients. Thus, our models seem more attuned to male patients in mixed-sex training sets, irrespective of the percentage of female patients. The best results are achieved when both sexes are trained exclusively on their respective data. Despite this, a male subgroup bias is apparent as the results for female patients are significantly worse than for male patients when trained exclusively on their data.

Base Model Reveals Sex Bias. The base model shows a significant sex bias in performance. When only male patients are involved the base model reveals a substantial performance gap between male and female patients. The results are significantly worse for female patients in male-skewed scenarios, except for the female-only experiments. Interestingly, the base model performs better for female patients than the adversarial and reinforcing models in the F75M25 dataset scenario. It is worth noting that the performance gap between the subgroups is not as apparent when the dataset includes male and female patients, unlike in the experiment that only involved male patients.

Reinforcement Model has No Significant Effect on Sex Bias. Training the model only on male patients increases AUC score variability and reduces performance differences between both sexes. The reinforcement model does not significantly affect sex bias.

Adversarial Model Eliminates Sex Bias in Cases Involving Only Female Patients. The adversarial model reduces sex bias significantly in scenarios with only female patients but is less effective in other scenarios, often favouring male patients. Its performance varies across experiments and datasets.

4   Discussion and Conclusions

We studied model and subgroup performance across datasets to identify the influence of bias in datasets on fairness of model predictions. We used linear programming (LP) to create various datasets with controlled male-female ratios. This was done to systematically evaluate the performance of three different learning strategies using ResNet-based CNNs. Other fairness and bias studies that require a flexible method to create datasets under certain constraints could benefit from this universal LP technique.

Our study shows that eliminating bias is challenging. The adversarial model architecture is able to reduce sex bias in a female-only context but fails for other datasets. Other model approaches do not show convincing results with respect to bias reduction.

Skewed sex distributions still show a performance gap between male and female patients. Our experiments demonstrate that the adversarial model better corrects sex bias in female-only datasets and not in male-only datasets, possibly due to other confounding and/or unidentified factors. Further research is needed on this issue.

As expected the base model shows a sensitivity for sex bias, possibly due to overfitting. The reinforcing and adversarial models both having a form of regularisation (to counter overfitting), are potentially able to reduce sex bias compared to the base model. However in our experiments we only see a bias correction for adversarial models for female-only experiments.

Our outcomes show that sex-related information influences prediction tasks. Future research should determine which specific sex-related factors are essential to ensure fairness across different subgroups.

In contrast to categorical data like patient sex, where the groups are clearly defined, this is not possible or only partially possible with continuous data like age, which could lead to somewhat arbitrary subgroups. Therefore, we started with the demographic attribute sex and will continue similar research with the non-categorical age attribute.

Further, we have identified the following directions for future work:

- Investigate whether using "early stopping" per task in a multi-task model reduces subgroup bias.
- Explore the impact of integrating segmentation with a classifier on sex-based disparities in identifying skin lesions.
- Study the roles of factors like skin colour and image artefacts in model fairness for different subgroups.
- Investigate the impact of shortcut learning on model fairness.

In conclusion, while we progress towards fairness, further advancements are needed to ensure consistent and equitable performance across various data distributions.

Acknowledgments. We gratefully acknowledge financial support from the Netherlands Organization for Scientific Research (NWO), grant no. 023.014.010.

Disclosure of Interests. The authors declare that they have no known competing financial interests or personal relationships that could have influenced the work reported in this paper.

References

1. Abbasi-Sureshjani, S., Raumanns, R., Michels, B.E., Schouten, G., Cheplygina, V.: Risk of training diagnostic algorithms on data with demographic bias. In: MICCAI LABELS Workshop, LNCS, vol. 12446, pp. 183–192. Springer (2020). https://doi.org/10.1007/978-3-030-61166-8_20
2. Adeli, E., et al.: Representation learning with statistical independence to mitigate bias. IEEE Winter Conf. Appl. Comput. Vis. **2021**, 2512–2522 (2021)
3. Bejnordi, B.E., et al.: Diagnostic assessment of deep learning algorithms for detection of lymph node metastases in women with breast cancer. JAMA **318**(22), 2199–2210 (2017)
4. Benčević, M., Habijan, M., Galić, I., Babin, D., Pižurica, A.: Understanding skin color bias in deep learning-based skin lesion segmentation. Comput. Methods Programs Biomed. **245**, 108044 (2024)
5. Bevan, P.J., Atapour-Abarghouei, A.: Skin deep unlearning: artefact and instrument debiasing in the context of melanoma classification. arXiv preprint arXiv:2109.09818 (Apr 2023)
6. Bissoto, A., Valle, E., Avila, S.: Debiasing skin lesion datasets and models? Not so fast (2020)
7. Cassidy, B., Kendrick, C., Brodzicki, A., Jaworek-Korjakowska, J., Yap, M.H.: Analysis of the ISIC image datasets: usage, benchmarks and recommendations. Med. Image Anal. **75**, 102305 (2022)

8. Chu, Z., Rathbun, S.L., Li, S.: Multi-Task adversarial learning for treatment effect estimation in basket trials. In: Flores, G., Chen, G.H., Pollard, T., Ho, J.C., Naumann, T. (eds.) Proceedings of the Conference on Health, Inference, and Learning. Proceedings of Machine Learning Research, vol. 174, pp. 79–91. PMLR (2022)

9. Codella, N., et al.: Skin lesion analysis toward melanoma detection 2018: a challenge hosted by the international skin imaging collaboration (ISIC) (2019)

10. Codella, N.C.F., et al.: Skin lesion analysis toward melanoma detection: a challenge at the 2017 international symposium on biomedical imaging (ISBI), hosted by the international skin imaging collaboration (ISIC) (2018)

11. Combalia, M., et al.: BCN20000: dermoscopic lesions in the wild (2019)

12. Esteva, A., et al.: Dermatologist-level classification of skin cancer with deep neural networks. Nature **542**(7639), 115–118 (2017)

13. Geirhos, R., et al.: Shortcut learning in deep neural networks. Nat. Mach. Intell. **2**(11), 665–673 (2020)

14. Géron, A.: Hands-On Machine Learning with Scikit-Learn, Keras, and TensorFlow. "O'Reilly Media, Inc." (Oct 2022)

15. Gichoya, J.W., et al.: AI recognition of patient race in medical imaging: a modelling study. Lancet Digit. Health **4**(6), e406–e414 (2022)

16. Groh, M., Harris, C., Daneshjou, R., Badri, O., Koochek, A.: Towards transparency in dermatology image datasets with skin tone annotations by experts, crowds, and an algorithm. Proc. ACM Hum.-Comput. Interact. **6**(CSCW2), 1–26 (2022)

17. Groh, M., et al.: Evaluating deep neural networks trained on clinical images in dermatology with the fitzpatrick 17k dataset. In: 2021 IEEE/CVF Conference on Computer Vision and Pattern Recognition Workshops (CVPRW), pp. 1820–1828 (Apr 2021)

18. Gutman, D., et al.: Skin lesion analysis toward melanoma detection: a challenge at the international symposium on biomedical imaging (ISBI) 2016, hosted by the international skin imaging collaboration (ISIC) (2016)

19. He, K., Zhang, X., Ren, S., Sun, J.: Deep residual learning for image recognition. In: Proceedings of the IEEE Conference on Computer Vision and Pattern Recognition, pp. 770–778 (2016)

20. ISIC archive. https://gallery.isic-archive.com. Accessed 7 June 2024

21. Jiménez-Sánchez, A., Juodelyte, D., Chamberlain, B., Cheplygina, V.: Detecting shortcuts in medical images-a case study in chest x-rays. In: 2023 IEEE 20th International Symposium on Biomedical Imaging (ISBI), pp. 1–5. IEEE (2023)

22. Larrazabal, A.J., Nieto, N., Peterson, V., Milone, D.H., Ferrante, E.: Gender imbalance in medical imaging datasets produces biased classifiers for computer-aided diagnosis. Proc. Natl. Acad. Sci. **117**(23), 12592–12594 (2020)

23. Liu, X., Shi, J., Zhou, S., Lu, M.: An iterated Laplacian based semi-supervised dimensionality reduction for classification of breast cancer on ultrasound images. In: International Conference of the IEEE Engineering in Medicine and Biology Society, vol. 2014, pp. 4679–4682 (2014). https://doi.org/10.1109/EMBC.2014.6944668

24. Marques, S., Schiavo, F., Ferreira, C.A., Pedrosa, J., Cunha, A., Campilho, A.: A multi-task CNN approach for lung nodule malignancy classification and characterization. Expert Syst. Appl. **184**, 115469 (2021)

25. Nauta, M., Walsh, R., Dubowski, A., Seifert, C.: Uncovering and correcting shortcut learning in machine learning models for skin cancer diagnosis. Diagnostics (Basel) **12**(1), 40 (2021)

26. Paszke, A., et al.: Others: an imperative style, high-performance deep learning library. Adv. Neural. Inf. Process. Syst. **32**, 8026–8037 (2019)

27. Petersen, E., et al.: Feature robustness and sex differences in medical imaging: a case study in MRI-based alzheimer's disease detection. In: International Conference on Medical Image Computing and Computer-Assisted Intervention, pp. 88–98. Springer (2022). https://doi.org/10.1007/978-3-031-16431-6_9
28. Raumanns, R., Schouten, G., Joosten, M., Pluim, J.P.W., Cheplygina, V.: Enhance (enriching health data by annotations of crowd and experts): a case study for skin lesion classification. Machine Learning for Biomedical Imaging **1**, 1–26 (2021). https://doi.org/10.59275/j.melba.2021-geb9, https://melba-journal.org/2021:020
29. Rotemberg, V., et al.: A patient-centric dataset of images and metadata for identifying melanomas using clinical context. Scientific Data; London **8**(1), s41597–021 (2021)
30. Saha, A., et al.: Artificial intelligence and radiologists in prostate cancer detection on MRI (PI-CAI): an international, paired, non-inferiority, confirmatory study. Lancet Oncol. (2024)
31. Seth, P., Pai, A.K.: Does the fairness of your Pre-Training hold up? Examining the influence of Pre-Training techniques on skin tone bias in skin lesion classification. In: Proceedings of the IEEE/CVF Winter Conference on Applications of Computer Vision, pp. 570–577 (2024)
32. Tschandl, P., Rosendahl, C., Kittler, H.: The HAM10000 dataset, a large collection of multi-source dermatoscopic images of common pigmented skin lesions (2018)
33. Willemink, M.J., et al.: Preparing medical imaging data for machine learning. Radiology **295**(1), 4–15 (2020)
34. Wu, Y., Zeng, D., Xu, X., Shi, Y., Hu, J.: FairPrune: achieving fairness through pruning for dermatological disease diagnosis. In: Medical Image Computing and Computer Assisted Intervention – MICCAI 2022, pp. 743–753. Springer Nature Switzerland (2022). https://doi.org/10.1007/978-3-031-16431-6_70
35. Yang, J., Soltan, A.A.S., Eyre, D.W., Yang, Y., Clifton, D.A.: An adversarial training framework for mitigating algorithmic biases in clinical machine learning. NPJ Digit. Med. **6**(1), 55 (2023)

AI Fairness in Medical Imaging: Controlling for Disease Severity

Pritam Mukherjee[ORCID] and Ronald M. Summers[✉][ORCID]

National Institutes of Health Clinical Center, 10 Center Dr, Bethesda,
MD 20892-1182, USA
{pritam.mukherjee,rms}@nih.gov

Abstract. A new criterion for assessing fairness of AI models in medical imaging is proposed. The key idea is to control for disease severity, which as a mediator, affects the presentation of disease in medical images, and hence the performance of AI algorithms. Existing fairness criteria such as equalized odds do not capture this effect, as is illustrated by an example. Additionally, a new metric is proposed based on the information theoretic notion of adjusted mutual information. The metric is easy to compute, captures the overall bias of an AI model, and can be used to compare multiple models, which is illustrated using an example. In this example, three chest X-ray classification models, trained on NIH, CheXpert and PadChest datasets, respectively, are used to predict on a subset of MIMIC-CXR cases, for which the severity scores of pulmonary edema are available, and the bias of the three models are computed and compared along two sensitive attributes: sex and ethnicity.

Keywords: Fairness · severity · AI · bias

1 Introduction

With the proliferation of artificial intelligence (AI) models that are increasingly being approved by regulatory authorities and put in clinical practice, the question of AI fairness has become salient in recent years. The realm of medical imaging, and particularly radiology, is no exception. AI models have been developed for a plethora of image based tasks – from identifying findings on chest X-rays [10,11,19] to automated segmentation of organs [12,16,20] to detection and segmentation of lesions and abnormalities in X-ray, computed tomography (CT), magnetic resonance imaging (MRI), ultrasound (US) and natural images [1]. Naturally, questions have arisen regarding the fairness of these models, since using a biased model in the clinic may not only harm individuals but also exacerbate existing inequities in the healthcare system.

Deep learning-based computer vision models may make assessment and enforcement of fairness particularly difficult, since they can learn subtle cues from the images about sensitive attributes such as race that are not easily evident to even experienced humans. For example, deep learning models can be

E. Puyol-Antón et al. (Eds.): FAIMI 2024/EPIMI 2024, LNCS 15198, pp. 24–33, 2025.
https://doi.org/10.1007/978-3-031-72787-0_3

trained to detect race from chest X-rays [5] very accurately – a task that is usually neither of interest, nor easy to do for human radiologists. In fact, despite many experiments with altering the images, the authors of [5] were unable to pinpoint the exact features the deep learning model was utilizing to determine race so accurately. The natural question then is whether these models are biased with respect to race when diagnosing disease or identifying abnormal findings.

Recent work has shown there is an empirical underdiagnosis bias for most of the existing deep learning models trained to identify findings on chest X-rays. In [17], the models were more likely to miss findings for underserved communities such as Blacks and women. However, it was not clear if the observed bias stemmed from external factors such as disease prevalence or age distribution which can affect the performance of AI algorithms and skew the analysis of bias [2,14]. Later work [6] confirmed subgroup disparities for chest X-ray models, but also found that correcting for some of the prevalence and population shifts partially mitigated the observed biases. Similar instances of empirical bias has also been observed for large image and text-based foundation models [7,15].

These prior works underscore the importance of assessing fairness for developed AI models. Crucially, the assessment of fairness is a non-trivial endeavour, mired in difficulties such as disentangling the bias that stems from the trained AI model itself from the bias introduced by external factors such as shifts in prevalence of diseases and age distribution. An intuitive notion of fairness requires that images with "similar" presentations of a disease should have "similar" outcomes. The key practical question, of course, is how to measure the similarity of two images, since the measure itself should not depend on the sensitive attribute. For example, a similarity metric which deems that two images are similar if and only if they come from a particular subgroup, would not be desirable. Many factors such as the co-occurrence of multiple findings, severity of disease, and scan quality can affect the presentation of a disease in a medical image. In this paper, we posit that the severity of the disease is a key factor, and that it should be controlled for when assessing bias of AI models.

Thus, the main contributions of our work are: (1) We propose a new fairness criterion which accounts for disease severity when assessing bias of AI models, and show its relation to the widely used "equalized odds" criterion (2) We demonstrate the inadequacy of the equalized odds criterion using a synthetic dataset, (3) We propose a related metric based on mutual information, which can be used to quantitatively measure bias and compare fairness between models, and (4) We assess and compare the fairness of three existing chest X-ray classification models on a real world pulmonary edema dataset.

2   Problem Formulation

Let P be an individual drawn randomly from an underlying population $\mathcal{P}$. The individual P is characterized by the tuple (X, Y, G), where $X \in \mathcal{X}$ denotes a medical image (such as a chest X-ray or CT scan), $Y \in \mathcal{Y}$ denotes the target variable or outcome of interest (for example, the presence or absence of a finding on a chest X-ray), and $G \in \mathcal{G}$ denotes a sensitive attribute (such as sex or race) for P. Our objective is to assess the fairness of a model $\mathcal{M}$ across the subgroups defined by G (for example, male vs female, White vs Black, etc.) on P. $\mathcal{M}$ takes as input the image X and tries to predict Y; its output is denoted by $\hat{Y} \in \hat{\mathcal{Y}}$.

In many applications of AI to medical imaging, the outcome Y is related to some disease (for example, detecting, segmenting, or classifying findings or diseases). Let $S \in \mathcal{S}$ denote the severity of the disease, and let us augment the tuple representing P to (X, Y, G, S). We propose the following fairness criterion.

2.1   A Fairness Criterion

Definition 1. A model $\mathcal{M}$ is fair if the following condition holds:

$$\mathbb{P}(\hat{Y}|S = s, G = g) = \mathbb{P}(\hat{Y}|S = s, G = g') \quad \forall s \in \mathcal{S}, \, g, g' \in \mathcal{G} \tag{1}$$

Heuristically, the condition states that the conditional distribution of the predictions given the severity of the disease is the same across all subgroups. Note that $\hat{Y} \in \hat{\mathcal{Y}}$ can be modeled as either discrete (e.g., binary outputs) or continuous (e.g., logits), depending on the application, in the above definition.

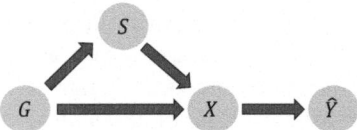

Fig. 1. Disease severity (S) is a mediator in the causal graph from the group variable G to the model prediction $\hat{Y}$.

Motivation: The presentation of a disease on a medical image, and hence the performance of algorithms (or humans) to detect or classify diseases can be strongly affected by the severity of the disease. For example, the size of a lesion affects the detection performance of both AI algorithms and humans – large lesions are usually easier to identify than small lesions. Formally, the severity of disease S is a mediator variable, providing an indirect path from the group variable G to the prediction $\hat{Y}$ (see Fig. 1), and therefore, one has to control for S when trying to discern the effect of G on $\hat{Y}$. The path from G to S may result from various reasons that are extrinsic to the model $\mathcal{M}$, including the nature of the subgroups (for example, if age is the group variable, chronic diseases may present with more severity in older individuals) or due to social inequities such as access to healthcare (for example, underserved communities may have less access to healthcare and may get imaging only at a more severe stage of a disease).

The proposed notion is closely related to the "equalized odds" group fairness criterion which requires:

$$\mathbb{P}(\hat{Y}|Y = y, G = g) = \mathbb{P}(\hat{Y}|Y = y, G = g') \quad \forall y \in \mathcal{Y}, \, g, g' \in \mathcal{G} \tag{2}$$

Thus, if $Y = S$ in some setting, the criterion proposed in Definition 1 will be equivalent to the equalized odds criterion. In general however, these two criteria are not equivalent and can, in fact, be mutually exclusive in many practical applications. For example, consider the common application where Y is binary and represents the presence or absence of a disease. In this case, Y can be a deterministic function of S, for example, $Y = S = 0$ if the disease is absent, while $Y = (S > 0)$ (assuming higher severity scores represent higher severity), if the disease is present. In such a formulation, we have

Theorem 1. *Assume $Y \in \{0, 1\}$ is binary. Further assume that $S = 0$ iff $Y = 0$, and $Y = S > 0$, where $S \in \{0, \ldots, S_{max}\}$, with $S_{max} > 1$. Finally, define $\alpha_s(g)$:*

$$\alpha_s(g) \triangleq \frac{\mathbb{P}(S = s | G = g)}{\mathbb{P}(Y = 1 | G = g)} \tag{3}$$

representing the fraction of individuals with severity s among the disease positive individuals of group g, and assume that the vector $[\alpha_1(g), \ldots, \alpha_{S_{max}}(g)]$ is drawn from some (unknown) continuous distribution on the S-dimensional probability simplex for every $g \in \mathcal{G}$. Then, the equalized odds criterion and the fairness criterion proposed in Definition 1 are mutually exclusive with probability 1.

Proof. Assume both equations (1) and (2) hold. Let

$$\mathbb{P}(\hat{Y} | S = s, G = g) \triangleq P_s(\hat{Y}), \quad \mathbb{P}(\hat{Y} | Y = 1, G = g) \triangleq Q(\hat{Y}) \quad \forall g \in \mathcal{G} \tag{4}$$

where $P_s(\hat{Y})$ and $Q(\hat{Y})$ do not depend on g due to (1) and (2), respectively. Then, we have,

$$Q(\hat{Y}) \triangleq \mathbb{P}(\hat{Y} | Y = 1, G = g) = \frac{\mathbb{P}(\hat{Y}, Y = 1 | G = g)}{\mathbb{P}(Y = 1 | G = g)} \tag{5}$$

$$= \frac{\sum_{s=1}^{S_{max}} \mathbb{P}(\hat{Y}, S = s | G = g)}{\mathbb{P}(Y = 1 | G = g)} \tag{6}$$

$$= \sum_{s=1}^{S_{max}} \alpha_s(g) P_s(\hat{Y}) \tag{7}$$

Note that equation (7) is satisfied for all $g \in \mathcal{G}$, and all $\hat{y} \in \hat{\mathcal{Y}}$ if either $P_s(\hat{Y})$ is the same for all s (meaning, the distribution of the outputs is independent of disease severity), or if $\alpha_s(g)$ does not depend on g for all s. If neither condition holds, we can rewrite equation (7) as

$$\sum_{s=1}^{S_{max}} (\alpha_s(g) - \alpha_s(g')) P_s(\hat{y}) = 0 \quad \forall g, g' \in \mathcal{G}, \hat{y} \in \hat{\mathcal{Y}} \tag{8}$$

In other words, $(\alpha_s(g) - \alpha_s(g'))$ must lie in the null space of the $|\hat{\mathcal{Y}}| \times |\mathcal{S}|$ matrix A with $A_{ij} = P_j(i)$. Given the distribution assumption on $\alpha_s(g)$, this does not hold with probability 1.

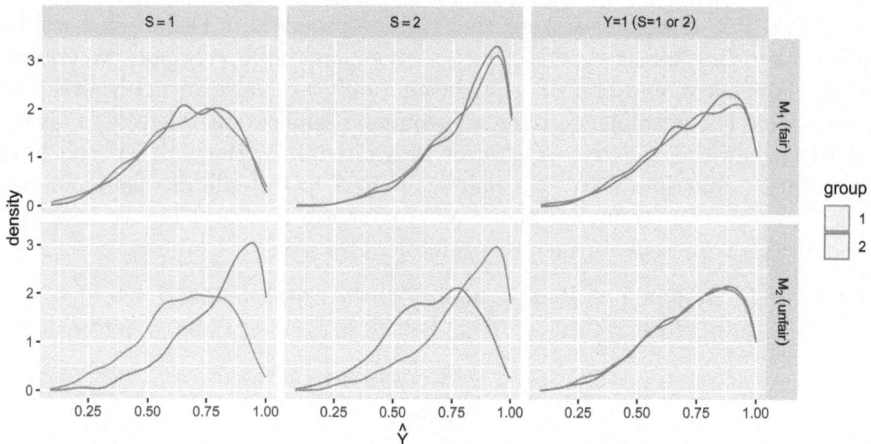

Fig. 2. An example where an unfair algorithm can satisfy the equalized odds criterion. Both model $\mathcal{M}_1$ and $\mathcal{M}_2$ are fair w.r.t. the equalized odds criterion (third column showing distribution of $\hat{Y}$ is similar for both groups), but model $\mathcal{M}_2$ is clearly unfair when the predictions are stratified by disease severity.

An Example on the Inadequacy of the Equalized Odds Criterion. To show the need for the proposed criterion, we construct an example to show that models can be inherently unfair despite satisfying the equalized odds criterion. To do so, consider the following setup:

$$\mathcal{G} = \{1, 2\}, \quad \mathcal{S} = \{0, 1, 2\}, \quad \mathcal{Y} = \{0, 1\}, \quad \alpha_s(g) = \frac{1}{2} \quad \forall s \in \mathcal{S}, g \in \mathcal{G} \quad (9)$$

Assume $S = 0$ iff $Y = 0$, and $Y = S > 0$. Consider a model $\mathcal{M}_1$ satisfying:

$$\mathbb{P}(\hat{Y}|S = 0, G = 1) = \mathbb{P}(\hat{Y}|S = 0, G = 2) = P \quad (10)$$

$$\mathbb{P}(\hat{Y}|S = 1, G = 1) = \mathbb{P}(\hat{Y}|S = 1, G = 2) = Q \quad (11)$$

$$\mathbb{P}(\hat{Y}|S = 2, G = 1) = \mathbb{P}(\hat{Y}|S = 2, G = 2) = R \quad (12)$$

and a model $\mathcal{M}_2$ satisfying:

$$\mathbb{P}(\hat{Y}|S = 0, G = 1) = \mathbb{P}(\hat{Y}|S = 0, G = 2) = P \quad (13)$$

$$\mathbb{P}(\hat{Y}|S = 1, G = 1) = \mathbb{P}(\hat{Y}|S = 2, G = 2) = Q \quad (14)$$

$$\mathbb{P}(\hat{Y}|S = 1, G = 2) = \mathbb{P}(\hat{Y}|S = 2, G = 1) = R \quad (15)$$

where $Q \neq R$. It is clear that $\mathcal{M}_1$ satisfies the proposed fairness criterion in equation (1), while $\mathcal{M}_2$ does not. However, it is easy to verify using (10), (13), and (7) that both $\mathcal{M}_1$ and $\mathcal{M}_2$ satisfy

$$\mathbb{P}(\hat{Y}|Y = 0, G = 1) = \mathbb{P}(\hat{Y}|Y = 0, G = 2) = P \quad (16)$$

$$\mathbb{P}(\hat{Y}|Y = 1, G = 1) = \mathbb{P}(\hat{Y}|Y = 1, G = 2) = \frac{1}{2}(Q + R) \quad (17)$$

In other words, both $\mathcal{M}_1$ and $\mathcal{M}_2$ would be considered fair under a equalized odds criterion. As a concrete example, Fig. 2 shows the distributions of the predictions when we choose $Q \sim \text{Beta}(4, 2)$ and $R \sim \text{Beta}(4, 1)$.

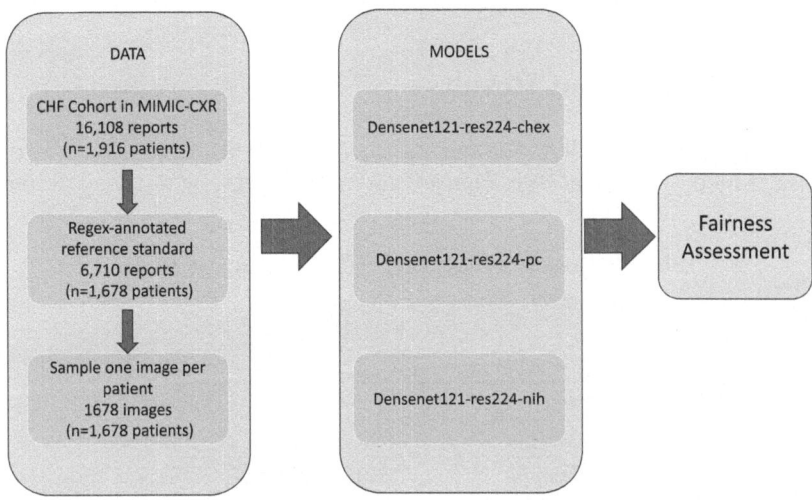

Fig. 3. Overview of the experiment.

2.2 Measuring Fairness

While we have established a criterion for fairness which accounts for disease severity, we would like to compare the degree to which two or more models are unfair – a critical task since practical models are unlikely to be perfectly fair. We can use the following information theoretic metric for fairness $\mathcal{F}$:

$$\mathcal{F}(\mathcal{M}) = I(\hat{Y}; G|S) = \sum_s \mathbb{P}(S = s)I(\hat{Y}; G|S = s) \qquad (18)$$

where $I(\cdot)$ represents mutual information. Note that $\mathcal{F}(\mathcal{M}) = 0$ iff $\mathcal{M}$ is perfectly fair, and $\mathcal{F}(\mathcal{M}) > 0$, otherwise. Since $\hat{Y}$ is typically a continuous random variable, computing this in practice may be non-trivial. Additionally, mutual information is not normalized and tends to increase in value with the number of categories in G simply due to chance. Therefore, we suggest the following alternative using adjusted mutual information (AMI) [18]:

$$\mathcal{F}(\mathcal{M}) = \sum_s \mathbb{P}(S = s) \left[\max_t |\text{AMI}(\hat{Y}_t; G|S = s)| \right] \qquad (19)$$

where $\hat{Y}_t = \hat{Y} > t$ is a binarized version of $\hat{Y}$ with threshold t. This is motivated by the fact that $\max_t I(\hat{Y}_t; G|S)$ is a lower bound for $I(\hat{Y}; G|S)$. Also, by using the absolute values of AMI, $0 \leq \mathcal{F}(\mathcal{M}) \leq 1$. Given n observations of $\hat{Y}$, one needs to consider $(n-1)$ thresholds, since any threshold picked within two consecutive values of the sorted $\hat{Y}$ leads to the same binarization. Finally, note that the metric depends on the distribution of severity in the test dataset, and cannot be used to compare models tested on different datasets.

3   Experiments and Results

In this section, we show how the fairness metric defined in (19) can assess the fairness of models on a dataset. Figure 3 shows the overview of the experiment.

Table 1. Demographics of the PE cohort ($n = 1,678$).

	Severity			
	0	1	2	3
Sex				
Male	354	176	192	101
Female	338	193	257	64
Unknown	2	1	0	0
Ethnicity				
White	456	225	315	117
Black	163	105	78	29
Hispanic/Latino	27	13	18	4
Asian	25	9	10	4
Unknown	23	17	27	9

Data. We used a publicly available dataset: "Pulmonary Edema Severity Grades Based on MIMIC-CXR" [9,13], henceforth referred to as PE, which assigned severity scores to a subset of chest X-ray images in the MIMIC-CXR dataset [8,11]. Briefly, the dataset comprised 17,857 images from 1,916 patients in MIMIC-CXR with an emergency department discharge diagnosis code consistent with congestive heart failure (CHF). Using regular expression matching, edema severity grades were extracted from radiology reports and numerically coded as follows: 0, none; 1, vascular congestion; 2, interstitial edema; and 3, alveolar edema. We use these as the reference standard in this work. Finally, to avoid multiple correlated images from the same patient in the test dataset, we kept only one image per patient. The demographics of the test patients are in Table 1.

Models. We evaluate three chest X-ray models on the PE dataset: one trained on the CheXpert [10], one trained on the PadChest [3], and another trained on the NIH ChestX-ray14 [19] dataset. To ensure reproducibility, we used the corresponding pretrained models "densenet121-res224-chex", "densenet121-res224-pc", and "densenet121-res224-nih", available from thee TorchXRayVision library [4]. Each of these models uses a Densenet-121 architecture, resizes the chest X-ray images to 224×224 pixel dimensions, and predicts the presence or absence of edema in the image.

Fairness Results. We assessed fairness of the three models along two sensitive attributes: sex and ethnicity. We used the "adjusted_mutual_info_score" metric in scikit-learn package to compute the AMI. For measuring fairness using the equalized odds criterion, we use the same metric in equation (19), simply replacing S by $Y(= S > 0)$. The results are summarized in Table 2. While the NIH and CheXpert models seem similar in terms of fairness under an equalized odds criterion, we see a larger difference under the proposed metric which controls for severity. On the other hand, when ethnicity is the sensitive attribute, the NIH model seems more fair than the others both in terms of our proposed criterion as well as the equalized odds criterion. Note also that the trends for the two criteria are not necessarily consistent – w.r.t. ethnicity, the CheXpert model is more fair than the PadChest model under equalized odds, but the result is the opposite with our proposed criterion. The reasons behind the observed fairness gaps and ways to mitigate them are not immediately clear, but it is beyond the scope of the current work.

Table 2. Fairness metrics for the three models. NIH. CheX and PC refer the the models trained on NIH, CheXpert and the PadChest datasets, respectively. All numbers in percentage.

	NIH	CheX	PC
Sex			
Proposed	2.48	1.82	1.33
Equalized odds	1.62	1.64	0.90
Ethnicity			
Proposed	1.10	2.55	2.19
Equalized odds	0.84	1.26	1.38

4   Discussion and Conclusions

In this paper, we have argued that disease severity must be accounted for when assessing fairness of AI models. To that end, we have proposed a new criterion for fairness, as well as a new metric which captures the overall bias of a model and can be used to compare between multiple models. We showed the inadequacy of the existing equalized odds fairness criterion in the context of medical imaging using a synthetic example. Usual metrics such as differences in sensitivity can be inadequate in practice, since a drop in sensitivity for a subgroup is often accompanied by an increase of specificity or precision, and as such, a different metric such as area under the receiver operating characteristic curve (AUROC) metric may show no bias. The proposed metric, on the other hand, captures the overall association between model outputs and subgroups, and makes it easy to compare different models. With its nice properties such as nonnegativity, upper-bound of 1, and ease of computation, we believe it will be useful in practice.

Acknowledgments. This study was supported by the Intramural Research Program of the NIH Clinical Center, and used the computing resources of the Biowulf HPC cluster at the NIH.

References

1. Aggarwal, R., et al.: Diagnostic accuracy of deep learning in medical imaging: a systematic review and meta-analysis. NPJ Digit. Med. **4**(1), 65 (2021). https://doi.org/10.1038/s41746-021-00438-z
2. Bernhardt, M., Jones, C., Glocker, B.: Potential sources of dataset bias complicate investigation of underdiagnosis by machine learning algorithms. Nat. Med. **28**(6), 1157–1158 (2022). https://doi.org/10.1038/s41591-022-01846-8
3. Bustos, A., Pertusa, A., Salinas, J.M., de la Iglesia-VayÃÂą, M.: Padchest: a large chest X-ray image dataset with multi-label annotated reports. Med. Image Anal. **66**, 101797 (2020). https://doi.org/10.1016/j.media.2020.101797, https://www.sciencedirect.com/science/article/pii/S1361841520301614
4. Cohen, J.P., et al.: TorchXrayvision: a library of chest X-ray datasets and models (2022). https://proceedings.mlr.press/v172/cohen22a.html
5. Gichoya, J.W., et al.: AI recognition of patient race in medical imaging: a modelling study. Lancet Digit. Health **4**(6), e406–e414 (2022). https://doi.org/10.1016/s2589-7500(22)00063-2
6. Glocker, B., Jones, C., Bernhardt, M., Winzeck, S.: Algorithmic encoding of protected characteristics in chest X-ray disease detection models. eBioMedicine **89**, 104467 (2023). https://doi.org/10.1016/j.ebiom.2023.104467
7. Glocker, B., Jones, C., Roschewitz, M., Winzeck, S.: Risk of bias in chest radiography deep learning foundation models. Radiol. Artif. Intell. **5**(6), e230060 (2023). https://doi.org/10.1148/ryai.230060
8. Goldberger, A.L., et al.: Physiobank, Physiotoolkit, and Physionet: components of a new research resource for complex physiologic signals. Circulation **101**(23), E215–20 (2000). https://doi.org/10.1161/01.cir.101.23.e215, https://www.ncbi.nlm.nih.gov/pubmed/10851218, goldberger, A L Amaral, L A Glass, L Hausdorff, J M Ivanov, P C Mark, R G Mietus, J E Moody, G B Peng, C K Stanley, H E eng 2000/06/14 Circulation. **101**(23), e215-e220 (2000). https://doi.org/10.1161/01.cir.101.23.e215
9. Horng, S., Liao, R., Wang, X., Dalal, S., Golland, P., Berkowitz, S.J.: Deep learning to quantify pulmonary edema in chest radiographs. Radiol. Artif. Intell.

3(2), e190228 (2021). https://doi.org/10.1148/ryai.2021190228, https://doi.org/10.1148/ryai.2021190228, pMID: 33937857

10. Irvin, J., et al.: CheXpert: a large chest radiograph dataset with uncertainty labels and expert comparison. Proceedings of the AAAI Conference on Artificial Intelligence **33**(01), 590–597 (2019). https://doi.org/10.1609/aaai.v33i01.3301590

11. Johnson, A.E.W., et al.: MIMIC-CXR, a de-identified publicly available database of chest radiographs with free-text reports. Sci. Data **6**(1), 317 (2019). https://doi.org/10.1038/s41597-019-0322-0

12. Li, W., Yuille, A., Zhou, Z.: How Well Do Supervised Models Transfer to 3D Image Segmentation? In: The Twelfth International Conference on Learning Representations (2024)

13. Liao, R., Chauhan, G., Golland, P., Berkowitz, S., Horng, S.: Pulmonary edema severity grades based on MIMIC-CXR (2021)

14. Mukherjee, P., Shen, T.C., Liu, J., Mathai, T., Shafaat, O., Summers, R.M.: Confounding factors need to be accounted for in assessing bias by machine learning algorithms. Nat. Med. **28**(6), 1159–1160 (2022). https://doi.org/10.1038/s41591-022-01847-7

15. Omiye, J.A., Lester, J.C., Spichak, S., Rotemberg, V., Daneshjou, R.: Large language models propagate race-based medicine. NPJ Digit. Med. **6**(1), 195 (2023). https://doi.org/10.1038/s41746-023-00939-z

16. Qu, C., Zhang, T., Qiao, H., Tang, Y., Yuille, A.L., Zhou, Z., et al.: Abdomenatlas-8k: annotating 8,000 CT volumes for multi-organ segmentation in three weeks. Adv. Neural Inf. Process. Syst. **36** (2023)

17. Seyyed-Kalantari, L., Zhang, H., McDermott, M.B.A., Chen, I.Y., Ghassemi, M.: Underdiagnosis bias of artificial intelligence algorithms applied to chest radiographs in under-served patient populations. Nat. Med. **27**(12), 2176–2182 (2021). https://doi.org/10.1038/s41591-021-01595-0

18. Vinh, N.X., Epps, J., Bailey, J.: Information theoretic measures for clusterings comparison: variants, properties, normalization and correction for chance. J. Mach. Learn. Res. **11**(95), 2837–2854 (2010). http://jmlr.org/papers/v11/vinh10a.html

19. Wang, X., Peng, Y., Lu, L., Lu, Z., Bagheri, M., Summers, R.M.: ChestX-Ray8: hospital-scale chest X-Ray database and benchmarks on weakly-supervised classification and localization of common thorax diseases. In: 2017 IEEE Conference on Computer Vision and Pattern Recognition (CVPR). IEEE (2017). https://doi.org/10.1109/cvpr.2017.369

20. Wasserthal, J., et al.: TotalSegmentator: robust Segmentation of 104 Anatomic Structures in CT Images. Radiol. Artif. Intell. **5**(5), e230024 (2023). https://doi.org/10.1148/ryai.230024

Fair and Private CT Contrast Agent Detection

Philipp Kaess[1,2](✉) ⓘ, Alexander Ziller[1](✉) ⓘ, Lea Mantz[2,3] ⓘ,
Daniel Rueckert[1,4] ⓘ, Florian J. Fintelmann[2] ⓘ, and Georgios Kaissis[1,5] ⓘ

[1] AI in Healthcare and Medicine, Technical University of Munich, Munich, Germany
{philipp.kaess,alex.ziller}@tum.de
[2] Department of Radiology, Massachusetts General Hospital, Boston, USA
[3] Department of Diagnostic and Interventional Radiology, University Medical Center
of the Johannes Gutenberg University Mainz, Mainz, Germany
[4] Department of Computing, Imperial College London, London, UK
[5] Institute of Machine Learning in Biomedical Imaging, Helmholtz Munich, Germany

Abstract. Intravenous (IV) contrast agents are an established medical
tool to enhance the visibility of certain structures. However, their application substantially changes the appearance of Computed Tomography
(CT) images, which - if unknown - can significantly deteriorate the diagnostic performance of neural networks. Artificial Intelligence (AI) can
help to detect IV contrast, reducing the need for labour-intensive and
error-prone manual labeling. However, we demonstrate that automated
contrast detection can lead to discrimination against demographic subgroups. Moreover, it has been shown repeatedly that AI models can
leak private training data. In this work, we analyse the fairness of conventional and privacy-preserving AI models during the detection of IV
contrast on CT images. Specifically, we present models which are substantially fairer compared to a previously published baseline. For better
comparability, we extend existing metrics to quantify the fairness of a
model on a protected attribute in a single value. We provide a model,
fulfilling a strict Differential Privacy protection of $(\varepsilon, \delta) = (8, 2.8 \cdot 10^{-3})$,
which with an accuracy of 97.42% performs 5%-points better than the
baseline. Additionally, while confirming prior works, that strict privacy
preservation increases the discrimination against underrepresented subgroups, the proposed model is fairer than the baseline over all metrics
considering race and sex as protected attributes, which extends to age
for a more relaxed privacy guarantee.

Keywords: CT Contrast Detection · Fairness · Privacy

P. Kaess and A. Ziller—Equal contribution.

1   Introduction

Intravenous (IV) contrast agents are invaluable in the realm of medical imaging, particularly in enhancing the diagnostic capabilities of Computed Tomography (CT) scans. By introducing contrast media into the vascular system, clinicians can enhance the visibility of various anatomical structures and pathological conditions on medical images. However, for diagnostic AI models, the altered appearance of CT images due to contrast media presents a challenge. Since AI models typically exhibit limited flexibility, contrast-enhanced and non-contrast CT images must be analyzed differently - either by separately trained models or with adjusted parameters. Consequently, unknown, incomplete, or unreliable information regarding the presence of IV contrast agent in CT images can potentially compromise the accuracy and reliability of AI models [22, 28].

To address this challenge, AI-driven approaches are being developed to automatically detect the use of contrast material in CT images [30]. These approaches reduce the dependency on manual labelling, which is not only labour-intensive but also prone to errors [16]. In this paper, we present a fair and private AI model for the prediction of the presence of contrast agent in CT images. By providing guarantees for privacy and maintaining fairness among subgroups, we aim to improve the overall trustworthiness and usability of AI-assisted CT imaging.

Among several other crucial aspects of ethical AI, fairness and privacy are two of the most prominent [20]. In this work, we specifically analyse group fairness, implying that certain subgroups, defined by "protected attributes" such as age, sex or race, are all treated equally. Moreover, medical images contain a wealth of information, which gets encoded into neural networks during training. It has been shown repeatedly that these networks are susceptible to leaking information during or after training [3, 4, 6, 13, 15, 17]. The utilisation of Differential Privacy (DP) enables the introduction of a mathematical protection that guarantees an upper bound on the risk of data leakage. This risk is modulated by the concept of a privacy budget, which can be set according to the specific setting.

Prior works. Ye et al. [30] demonstrated the use of AI models for the detection of IV Contrast (IC) on CTs. We use their approach as a baseline to compare against our models. Several works have investigated the correlation between formal privacy preservation and the fairness of AI models. There is empirical [12, 14] and theoretical [9, 26] evidence that privacy preservation amplifies discrimination and biases in AI models. Specifically, it was shown that strict privacy preservation leads to an accuracy discrepancy between the overall model and an underrepresented group [26]. This discrepancy can be mitigated, however, for the cost of overall model accuracy. Hence, these fundamental requirements of ethical AI stand in conflict to each other. Other works have found that these biases might not only be driven by the (under-)representation of certain subgroups in the training data but by the diagnosis difficulty [29]. For example, they found that in the task of classifying chest x-rays of older patients, who are largely overrepresented yet harder to diagnose due to artifacts and other factors, strict privacy preservation disproportionately affects this group.

Background. The gold standard for providing mathematically provable privacy guarantees is Differential Privacy (DP). DP limits the contribution of individual patients to the final outcome and by that upper bounds the success of membership inference [10] and data reconstruction attacks [19]. Furthermore, it has been shown that DP can fulfil necessary conditions imposed by the EU GDPR [8]. The level of protection can be set depending on the specific requirements. In order to quantify the provided protection, the concept of a privacy budget expressed as a variable parameter ε and a fixed parameter δ is typically used. The most common method to provide such DP guarantees during the training of neural networks is DP-SGD, as e.g. defined by Abadi et al. [1]. Crucially, it depends on two main steps: (1) Clipping the per-sample gradients to a maximum gradient norm and (2) adding noise calibrated to the maximum gradient norm.

Terminology This paper deals with sensitive concepts such as sex and race. We acknowledge that these concepts have weaknesses as some individuals cannot identify themselves with any one category. Throughout this paper, when we talk about race, we are referring to the self-reported race that patients declared on admission to the hospital. Additionally, for the external test set we defined a separate race group *Hispanic* to also capture a possible discrimination against the group of people who self-identified to be of Hispanic ethnicity. Furthermore, sex denotes the legal sex a patient is associated with, although this might be in contrast to the self-identified gender.

2   Materials and Methods

2.1   Dataset

For the purpose of model development, we utilized a dataset ($n = 733$) comprising chest CT scans from patients who underwent lobectomy or bi-lobectomy at one of three institutions between July 1, 2014 and June 30, 2017 due to a presumed diagnosis of primary non-small cell lung cancer. These CT scans were obtained within 90 d prior to surgery. We will refer to this set as *Development Dataset*. It represents a subset from the multicenter body-composition study by Best et al. [2], selected based on the availability of reliable IV contrast labels within the original dataset. Each scan in this subset was manually annotated for the presence or absence of IC by a trained research assistant. For training, we performed a randomized 70/30 split of the Development Dataset into training/validation ($n = 513$) and internal test ($n = 220$) patient subsets.

For testing, we employed an external dataset ($n = 4\,478$), derived from all chest CT scans performed at Massachusetts General Hospital (MGH) between 1 January 2019 and 31 December 2019. This subset included only adult patients with known self-identified race and scans annotated by the attending radiologist as either with or without contrast agent in the radiology report. These IC labels, initially extracted from radiology reports, were subsequently manually

Table 1. Patient statistics of all used datasets.

	Training/Validation				Test
	Overall (n=733)	Center 1 (n=151)	Center 2 (n=390)	Center 3 (n=192)	External (n=4478)
With IC	441 (60.2%)	111 (73.5%)	231 (59.2%)	99 (51.6%)	2392 (53.4%)
Without IC	292 (39.8%)	40 (26.5%)	159 (40.8%)	93 (48.4%)	2086 (46.6%)
Female	411 (56.1%)	76 (50.3%)	234 (60.0%)	101 (52.6%)	2371 (52.9%)
Male	322 (43.9%)	75 (49.7%)	156 (40.0%)	91 (47.4%)	2107 (47.1%)
White	633 (86.4%)	104 (68.9%)	354 (90.8%)	175 (91.1%)	2317 (51.7%)
Black	62 (8.5%)	37 (24.5%)	11 (2.8%)	14 (7.3%)	637 (14.2%)
Asian	32 (4.4%)	10 (6.6%)	19 (4.9%)	3 (1.6%)	786 (17.6%)
Hispanic	–	–	–	–	226 (5.0%)
Other	3 (0.4%)	0 (0.0%)	3 (0.8%)	0 (0.0%)	512 (11.4%)
Unknown	3 (0.4%)	0 (0.0%)	3 (0.8%)	0 (0.0%)	–
Mean Age	67.2 ± 9.9	66.2 ± 9.6	68.4 ± 10.2	65.4 ± 9.1	63.7 ± 14.9

verified. Additionally, three non-contrast scans were excluded because annotators detected traces of contrast agent from previous procedures, making binary IC labeling impossible.

To enhance the fairness analysis, we increased the representation of racial minorities in the external test set by limiting data from White patients to the period from October 1, 2019, to December 31, 2019. This adjustment was applied only during prediction on the external test set and thus did not affect the model's training or fairness itself. Instead, it improved the statistical power for evaluating model performance on minority groups, allowing for a more meaningful fairness analysis. The Institutional Review Boards at all institutions involved approved this secondary analysis of a retrospective cohort study and waived the requirement for informed consent.

Patient statistics in Table 1 illustrate the artificially increased ratio of ethnic minorities: the *White* group comprises only 51.7% of the scans, compared to 86.4% in the Development Dataset, favoring minority groups. The differences in age, sex, and IC distribution between the datasets are due to the distinct patient cohorts. The development dataset includes only patients with lung cancer, which has a statistical influence on the demographic features, while the external test set covers a broader demographic, unified by the criterion of having undergone a CT scan at MGH within the specified timeframe. This cohort discrepancy likely accounts for the observed variations in demographic and clinical characteristics between the two datasets. Following the approach of Ye et al. [30] all CT images are clipped to be at least −1000 Hounsfield Units.

Table 2. Performance comparison of all models. The results of Ye et al. are retrieved by applying their published model to our test set. Our results are median test results over 10 trainings and their standard deviation. ε denotes the privacy budget. Lower values of ε correspond to stricter privacy guarantees. $\varepsilon = \infty$ corresponds to standard non-private AI training. MCC is the Matthew's Correlation Coefficient.

	ε $(\delta = \frac{1}{359})$	Accuracy	F1-Score	MCC
Ye et al. [30]	∞	92.82%	92.80%	86.58%
Ours	∞	99.53% $\pm$ 0.08%	99.56% $\pm$ 0.07%	99.06% $\pm$ 0.15%
	600	99.05% $\pm$ 0.89%	99.12% $\pm$ 0.80%	98.11% $\pm$ 1.73%
	37.5	96.99% $\pm$ 0.79%	97.25% $\pm$ 0.71%	94.05% $\pm$ 1.54%
	18.75	97.80% $\pm$ 0.93%	97.98% $\pm$ 0.82%	95.63% $\pm$ 1.79%
	8	97.42% $\pm$ 0.84%	97.64% $\pm$ 0.74%	94.90% $\pm$ 1.61%

2.2 Pipeline

Our preprocessing follows the approach of Ye et al. [30]. For classifying a CT scan we disassemble the 3D image into 75 2D axial slices, which we use as input to our model. We optimised the slice range, i.e. the slices used for the final prediction, on the (registered) 3D scan by minimising the overall mean squared error for all models on the validation set. Furthermore, we tuned the decision threshold of each model on the validation set. Our privacy guarantees are given over each CT scan. Assuming that each patient contributes exactly one scan, it also corresponds to a per-patient guarantee. In order to provide this guarantee, we calculate the gradient over all slices from one CT scan, average them and then perform standard clipping and noising as required by DP-SGD. For all private experiments, we set δ to be the inverse of the training set size, i.e., $\delta = \frac{1}{359}$. For all experiments, we used a ResNet-9 with ScaleNorm [21], which has proven to be efficient and robust both in private and non-private training settings. All private models were trained with a batch size of 16 3D CTs for 300 epochs. The learning rate starts at $4 \cdot 10^{-3}$ and decays with a factor of 0.5 every 100 epochs. We used a maximum gradient norm for clipping of 5 for $\varepsilon = 8$ and 20 for all others. The number of epochs, maximum gradient norm, and the learning rate were determined empirically on the validation set. As the non-private model converged substantially faster, it was trained for 150 epochs using a constant learning rate of $2 \cdot 10^{-3}$ and a batch size of 64 slices.

2.3 Fairness Metrics

There is no universally accepted definition of fairness in the evaluation of AI models, as evidenced by the variety of widely used fairness metrics, each measuring slightly different aspects. Prominent group fairness metrics include Equal Opportunity [18], Equalized Odds [18], Underdiagnosis Rate [27], Disparate Impact

(DI), and Statistical Parity Difference (SPD) [5, 11], which all represent (slightly) varying notions of fairness. In our problem setting, there is no possible under-diagnosis or immediately obvious favourable outcome as e.g. in the selection of job applicants. As neither a false negative nor a false positive is beneficial for a patient, the sole criterion to evaluate fairness is the correctness of a prediction, which we thus define as the *favourable outcome* for metrics calculations. Given these circumstances, we determined DI and SPD to be the most suitable con-tinuously measurable fairness metrics. In order to not only evaluate the impact on the specific subgroups but quantify how "fair" an entire model is, we extend these two metrics by calculating the variance over all subgroups as well as the minimal value of all subgroups. Specifically, we calculate as follows:

$$\text{SPD}(g) = P(\hat{Y} = 1 | g) - P(\hat{Y} = 1 | \bar{g}) \tag{1}$$

$$\text{DI}(g) = \frac{P(\hat{Y} = 1 | g)}{P(\hat{Y} = 1 | \bar{g})} \tag{2}$$

$$M_{\text{Var}} = \frac{1}{n-1} \sum_{g \in G} (M(g)) - \mu_{M_G})^2 \tag{3}$$

$$M_{\text{Min}} = \min_{g \in G} M(g) \tag{4}$$

where g are subgroups, $\bar{g}$ is the group of patients not in that subgroup, $M(g)$ is an arbitrary metric (SPD or DI) over subgroup g, μ_{M_G} the mean of this metric over all patients from all groups, and $P(\hat{Y} = 1 | i)$ denotes the probability of a correct model prediction (:= favourable outcome) for a patient in subgroup i. The variance of a fairness metric (SPD or DI) over all subgroups corresponds to a notion of fairness where *many* groups should have approximately the same outcome. The minimum metric is a type of min-max fairness. It only considers the DI or SPD value of the most discriminated subgroup and resembles a notion of fairness where *no* group should have a much lower chance of being correctly predicted. To prevent a possible influence of the choice of fairness metric on pos-sible conclusions we draw from our fairness analysis, we will utilize and compare all introduced metrics, i.e. SPD_{Var}, DI_{Var}, SPD_{Min} and DI_{Min}.

3   Results

In Table 2, we detail the performance of our models on various levels of privacy and compare it to the published model of Ye et al. [30]. For each model, we list the accuracy, F1-Score and Matthew's Correlation Coefficient (MCC) [7, 24] on the external test set. We find that even our most private models clearly outperform the published baseline. The median result of our most private model at $\varepsilon = 8$ drops by 2% in accuracy and 5% MCC compared to our non-private model but outperforms Ye et al. [30] by 5% in accuracy and 8% MCC.

Figure 1 visualizes the performance of all investigated models over all pro-tected attributes and privacy levels. For every privacy level, we trained ten sep-arate models on our development dataset, each initialized with different random

Table 3. Fairness evaluation of all models. SPD_{Var} and DI_{Var} denote the median $\pm$ standard deviation of the variance of statistical parity difference (SPD) and disparate impact (DI) over all subgroups. SPD_{Min} and DI_{Min} denote the median $\pm$ standard deviation of the SPD and DI of the most discriminated subgroup. $\downarrow$ denotes metrics where lower values are fairer, $\uparrow$ means larger values are fairer. The minimum metrics are measured in %, while the variances are provided in $\%^2$.

		ε	Race	Sex	Age
SPD_{Var} $\downarrow$	Ye et al. [30]	∞	97.67	42.61	13.66
	Ours	∞	0.98 ± 0.27	0.03 ± 0.12	0.20 ± 0.07
		600	1.11 ± 0.44	0.10 ± 0.49	1.62 ± 3.67
		37.5	1.23 ± 0.60	0.34 ± 0.49	7.58 ± 4.88
		18.75	1.39 ± 0.63	1.02 ± 3.12	2.85 ± 8.09
		8	1.70 ± 0.51	0.75 ± 0.67	10.29 ± 5.92
DI_{Var} $\downarrow$	Ye et al. [30]	∞	123.81	49.67	15.85
	Ours	∞	0.99 ± 0.27	0.03 ± 0.13	0.21 ± 0.07
		600	1.13 ± 0.45	0.10 ± 0.50	1.65 ± 3.91
		37.5	1.30 ± 0.63	0.36 ± 0.52	8.06 ± 5.24
		18.75	1.45 ± 0.69	1.11 ± 3.31	2.99 ± 8.97
		8	1.75 ± 0.53	0.79 ± 0.71	10.73 ± 6.51
SPD_{Min} $\uparrow$	Ye et al. [30]	∞	-9.37	-4.62	-5.06
	Ours	∞	-1.88 ± 0.25	-0.13 ± 0.13	-0.95 ± 0.23
		600	-1.83 ± 0.27	-0.22 ± 0.24	-2.92 ± 2.03
		37.5	-1.17 ± 0.63	-0.41 ± 0.25	-5.55 ± 2.46
		18.75	-1.81 ± 0.44	-0.71 ± 0.59	-3.53 ± 2.49
		8	-2.12 ± 0.57	-0.61 ± 0.34	-8.39 ± 2.10
DI_{Min} $\uparrow$	Ye et al. [30]	∞	90.05	95.14	94.57
	Ours	∞	98.11 ± 0.25	99.87 ± 0.13	99.05 ± 0.24
		600	98.16 ± 0.28	99.78 ± 0.24	97.05 ± 2.09
		37.5	98.80 ± 0.63	99.58 ± 0.25	94.27 ± 2.55
		18.75	98.15 ± 0.45	99.27 ± 0.60	96.38 ± 2.63
		8	97.85 ± 0.58	99.37 ± 0.35	91.43 ± 2.21

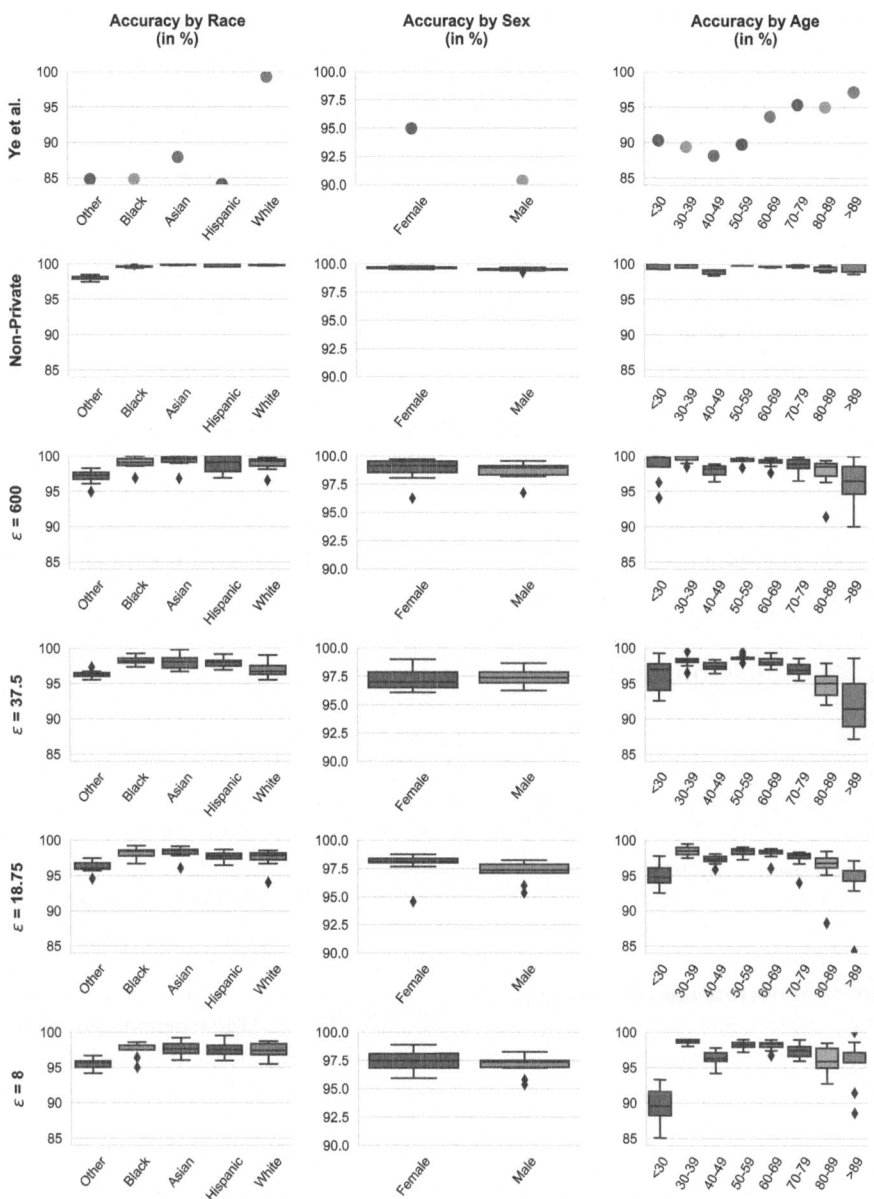

Fig. 1. Comparison of accuracies on the external test set across demographic subgroups for the model published by Ye et al. [30] (first row) and our models with varying levels of privacy (rows two to six). Each column shows the performance disparity of a specific protected attribute, namely, race, sex and age. Each of rows two through six presents the performance of ten models, each trained on our development dataset with (varying) random seeds, under the respective privacy budget. The median result is denoted as a line in the boxplot. Outlier results are marked as diamonds.

weights. We report their per-subgroup performance on the previously unseen external test set as a boxplot, where every box depicts the accuracies of the ten models with the respective privacy level on a specific demographic patient group. The fairness of each privacy setting is quantified by the median variance and minimum over SPD and DI in Table 3. We observe that our models over all privacy levels are substantially fairer on race and sex compared to the model published by Ye et al. [30], which in particular only performs well on patients of group *White*. Considering the age, the variance metrics indicate that the baseline is less fair than our models over all privacy levels, however, the minimum metrics for age of our most private model are worse than the baseline, indicating the importance of the choice of fairness metrics.

We also observe that with increasing protection of patient privacy models become slightly more unfair. For example, while our non-private model has a median DI_{Min} of 98.11% among the different racial subgroups, the most private model at $\varepsilon = 8$ exhibits a median DI_{Min} of 97.85%. Equivalently, the median SPD_{Var} increases from 0.98 for the non-private model to 1.70 for $\varepsilon = 8$. In comparison, the model published by Ye et al. [30] exhibits an SPD_{Var} of 97.67, implying an orders of magnitude larger variation in the classification performance across different racial subgroups groups.

4 Discussion and Conclusion

In this paper, we present a study of fair and privacy-preserving models for the detection of contrast agent in CT scans. Firstly, we note that our models are substantially fairer than an established baseline model. In clinical practice, this could mean that any non-white patients have substantially higher risks of being misclassified, leading to potential performance deterioration of downstream tasks that rely on correct contrast labels. This once more highlights the importance of reviewing AI models for any biases. Even though we did not have any specific modifications (such as regularisation or fairness metric optimisation terms) to our training pipeline, we could provide models which perform well independent of the patient's race, sex or age. Moreover, we could show that mathematically provable privacy guarantees are compatible with highly accurate and fair AI models. Although we experience the same effect as several prior works, that stronger privacy guarantees lead to increased biases, even our most private models are fairer than the baseline. The exact choice of an appropriate level of privacy depends on the specific use case and threat model, i.e. what attack and which capabilities of an attacker we aim to defend against. The privacy guarantees we studied in this work range from strict guarantees defending against Membership Inference Attacks of extremely powerful adversaries [25] to more relaxed privacy guarantees sufficient against reconstruction attacks of still powerful yet not omnipotent adversaries [31]. Furthermore, by measuring the variance and minimum of SPD and DI we extended these established metrics in a way such that they capture the fairness of an entire scenario. This allows the comparison of various models, such as in our case, over varying levels of privacy. The suitability for practical and legal contexts has to be further evaluated. In conclusion, we have presented a fair and private AI model for the task of CT contrast prediction. We view this as this is an important step towards the clinical use of such an AI model on real patients.

Acknowledgments. This work was supported by the German Ministry of Education and Research (BMBF) under grant number 01ZZ2316C (PrivateAIM). All models were trained using computational resources and services provided by the Massachusetts Life Sciences Center [23].

Disclosure of Interests. The authors declare no competing interests.

References

1. Abadi, M., et al.: Deep learning with differential privacy. In: Proceedings of the 2016 ACM SIGSAC Conference on Computer and Communications Security, pp. 308–318. ACM (2016). https://doi.org/10.1145/2976749.2978318

2. Best, T.D., et al.: Multilevel body composition analysis on chest computed tomography predicts hospital length of stay and complications after lobectomy for lung cancer: a multicenter study. Ann. Surg. **275**(5), e708–e715 (2022). https://doi.org/10.1097/SLA.0000000000004040. epub 2020 Jul 8 PMID: 32773626

3. Boenisch, F., Dziedzic, A., Schuster, R., Shamsabadi, A.S., Shumailov, I., Papernot, N.: When the curious abandon honesty: federated learning is not private. In: 2023 IEEE 8th European Symposium on Security and Privacy (EuroS&P), pp. 175–199. IEEE (2023)

4. Buzaglo, G., et al.: Deconstructing data reconstruction: multiclass, weight decay and general losses. In: Thirty-seventh Conference on Neural Information Processing Systems (2023)

5. Calders, T., Verwer, S.: Three naive Bayes approaches for discrimination-free classification. Data Min. Knowl. Disc. **21**(2), 277–292 (2010). https://doi.org/10.1007/s10618-010-0190-x

6. Carlini, N., et al.: Extracting training data from diffusion models. In: 32nd USENIX Security Symposium (USENIX Security 23), pp. 5253–5270 (2023)

7. Chicco, D., Jurman, G.: The advantages of the Matthews correlation coefficient (MCC) over F1 score and accuracy in binary classification evaluation. BMC Genomics **21**, 1–13 (2020)

8. Cohen, A., Nissim, K.: Towards formalizing the GDPR's notion of singling out. Proc. Natl. Acad. Sci. **117**(15), 8344–8352 (2020)

9. Cummings, R., Gupta, V., Kimpara, D., Morgenstern, J.: On the compatibility of privacy and fairness, pp. 309-315. UMAP'19 Adjunct, Association for Computing Machinery, New York, NY, USA (2019). https://doi.org/10.1145/3314183.3323847

10. Dong, J., Roth, A., Su, W.J.: Gaussian differential privacy. J. R. Stat. Soc. Ser. B Stat Methodol. **84**(1), 3–37 (2022)

11. Dwork, C., Hardt, M., Pitassi, T., Reingold, O., Zemel, R.: Fairness through awareness. In: Proceedings of the 3rd Innovations in Theoretical Computer Science Conference, pp. 214–226 (2012)

12. Farrand, T., Mireshghallah, F., Singh, S., Trask, A.: Neither private nor fair: impact of data imbalance on utility and fairness in differential privacy (2020)

13. Feng, S., Tramèr, F.: Privacy backdoors: stealing data with corrupted pretrained models. In: International Conference on Machine Learning. PMLR (2024)
14. Fioretto, F., Tran, C., Hentenryck, P.V.: Decision making with differential privacy under a fairness lens. In: International Joint Conference on Artificial Intelligence (2021). https://api.semanticscholar.org/CorpusID:234742410
15. Fowl, L., Geiping, J., Czaja, W., Goldblum, M., Goldstein, T.: Robbing the fed: directly obtaining private data in federated learning with modified models. In: Tenth International Conference on Learning Representations (2022)
16. Güld, M., et al.: Quality of DICOM header information for image categorization. In: Proceedings of SPIE - The International Society for Optical Engineering, vol. 4685 (2002). https://doi.org/10.1117/12.467017
17. Haim, N., Vardi, G., Yehudai, G., Shamir, O., Irani, M.: Reconstructing training data from trained neural networks. Adv. Neural. Inf. Process. Syst. **35**, 22911–22924 (2022)
18. Hardt, M., Price, E., Srebro, N.: Equality of opportunity in supervised learning. Adv. Neural Inf. Process. Syst. **29**, 3315–3323 (2016)
19. Hayes, J., Mahloujifar, S., Balle, B.: Bounding training data reconstruction in DP-SGD. In: Thirty-seventh Conference on Neural Information Processing Systems (2023)
20. Jobin, A., Ienca, M., Vayena, E.: The global landscape of AI ethics guidelines. Nat. Mach. Intell. **1**(9), 389–399 (2019)
21. Klause, H., Ziller, A., Rueckert, D., Hammernik, K., Kaissis, G.: Differentially private training of residual networks with scale normalisation. In: Theory and Practice of Differential Privacy Workshop, ICML (2022)
22. Lartaud, P.J., Rouchaud, A., Rouet, j.m., Nempont, O., Boussel, L.: Spectral CT Based Training Dataset Generation and Augmentation for Conventional CT Vascular Segmentation, pp. 768–775 (10 2019). https://doi.org/10.1007/978-3-030-32245-8_85
23. Massachusetts life sciences center: computational resources and services. https://www.masslifesciences.com/
24. Matthews, B.W.: Comparison of the predicted and observed secondary structure of t4 phage lysozyme. Biochim et Biophys. Acta (BBA)-Protein Structure **405**(2), 442–451 (1975)
25. Nasr, M., Songi, S., Thakurta, A., Papernot, N., Carlini, N.: Adversary instantiation: lower bounds for differentially private machine learning. In: 2021 IEEE Symposium on security and privacy (SP), pp. 866–882. IEEE (2021)
26. Sanyal, A., Hu, Y., Yang, F.: How unfair is private learning? In: Cussens, J., Zhang, K. (eds.) Proceedings of the Thirty-Eighth Conference on Uncertainty in Artificial Intelligence. Proceedings of Machine Learning Research, vol. 180, pp. 1738–1748. PMLR (8 2022). https://proceedings.mlr.press/v180/sanyal22a.html
27. Seyyed-Kalantari, L., Zhang, H., McDermott, M.B., Chen, I.Y., Ghassemi, M.: Underdiagnosis bias of artificial intelligence algorithms applied to chest radiographs in under-served patient populations. Nat. Med. **27**(12), 2176–2182 (2021)
28. Sofka, M., et al.: Automatic contrast phase estimation in CT volumes. In: Fichtinger, G., Martel, A., Peters, T. (eds.) Medical Image Computing and Computer-Assisted Intervention – MICCAI 2011, pp. 166–174. Springer Berlin Heidelberg, Berlin, Heidelberg (2011). https://doi.org/10.1007/978-3-642-23626-6_21
29. Tayebi Arasteh, S., et al.: Preserving fairness and diagnostic accuracy in private large-scale ai models for medical imaging. Commun. Med. **4**(1) (Mar

2024). https://doi.org/10.1038/s43856-024-00462-6. http://dx.doi.org/10.1038/s43856-024-00462-6

30. Ye, Z et al.: Deep learning-based detection of intravenous contrast enhancement on CT scans. Radiol. Artif. Intell. 4(3), e210285 (2022). https://doi.org/10.1148/ryai.210285

31. Ziller, A., et al.: Reconciling privacy and accuracy in AI for medical imaging. Nat. Mach. Intell. 1–11 (2024)

Mitigating Overdiagnosis Bias in CNN-Based Alzheimer's Disease Diagnosis for the Elderly

Vien Ngoc Dang[1]([envelope])[ID], Adrià Casamitjana[2][ID],
Jerónimo Hernández-González[3][ID], Karim Lekadir[1,4][ID],
and for the Alzheimer's Disease Neuroimaging Initiative

[1] Departament de Matemàtiques i Informàtica, Universitat de Barcelona, Barcelona,
Spain
dangn@ub.edu
[2] Research Institute of Computer Vision and Robotics, University of Girona, Girona,
Spain
[3] Departament d'Informàtica, Matemàtica Aplicada i Estadística, Universitat de
Girona, Girona, Spain
[4] Institució Catalana de Recerca i Estudis Avançats (ICREA), Barcelona, Spain

Abstract. Diagnosing Alzheimer's disease (AD) presents significant
challenges in the oldest populations due to overlapping symptoms of normal cognitive aging and early-stage dementia. While AI algorithms have
matched specialist performance in diagnosing AD, they tend to produce
unreliable results for the oldest populations, generating false positives
that increase radiologist workloads and healthcare costs. In this study,
we focus on mitigating overdiagnosis bias in CNN-based AD diagnosis
for these groups. We present a post-hoc bias mitigation technique that
significantly improves fairness by reducing overdiagnosis and enhances
reliability by improving calibration without compromising overall model
accuracy. Code is available at: https://github.com/ngoc-vien-dang/C-
GTOP.

Keywords: fairness · overdiagnosis bias · medical imaging ·
convolutional neural networks · Alzheimer's disease · elderly

1 Introduction

Diagnosing Alzheimer's disease (AD) presents significant challenges in the oldest populations. In clinical settings, this demographic group, typically those aged

Alzheimers Disease Neuroimaging Initiative—Data used in preparation of this article
was obtained from the Alzheimer's Disease Neuroimaging Initiative (ADNI) database
(http://www.adni-info.org/). The investigators within the ADNI contributed to the
design and implementation of ADNI and/or provided data, but did not participate in
analysis or writing of this report.

85 and older, is particularly prone to misdiagnosis due to the overlapping symptoms of normal cognitive aging and early-stage dementia. The natural cognitive decline associated with aging can be confounded with the early symptoms of AD, making accurate diagnosis difficult and subsequently leading to false positives. Additionally, image-based diagnosis is hindered by the presence of multiple comorbidities, brain resilience, and anatomical heterogeneity among elderly brains. While AI algorithms have demonstrated performance on par with specialists in diagnosing AD [16], their reliability in the oldest populations remains questionable. When tested in smaller studies about AD, deep learning models show large false positive rates [3], suggesting an algorithmic bias that mirrors the real-world diagnostic challenges that clinicians face at the extremes of age. A high rate of false positives would significantly harm the efficiency of screening processes, requiring additional resources to manage incorrect results.

Although substantial research has been conducted on algorithmic bias in AD diagnosis [3,12], the specific issue of AI-driven overdiagnosis in the oldest remains relatively unexplored. This gap in research is critical given the unique challenges this demographic group poses. Training with unbalanced data, with underrepresentation of cognitively normal (CN) older adults and a higher prevalence of AD, induces models biased towards overdiagnosing AD. Additionally, age-related anatomical changes affect per-age model calibration, leading to increased false positive rates, particularly affecting the oldest [3,6]. This discrimination in diagnostic performance calls for effective solutions. To address these disparities, previous research in healthcare has predominantly concentrated on group fairness [1,4,13], which aims to balance performance metrics across different protected groups. We evaluate the Calibrated Equalized Odds Post-Processing (CPP) [15] method as a representative for group fairness. To achieve group fairness, specifically in misdiagnosis rates across age groups, CPP increases the error rate for untreated groups (younger age groups) and exacerbates calibration bias, which adversely affects other subgroups. Our findings align with prior studies [4] demonstrating that while CPP can achieve performance parity, it often worsens performance across all groups. Consequently, it remains uncertain when this definition of fairness is suitable for clinical settings.

In this paper, we address the challenge of overdiagnosis bias in CNN-based AD diagnosis among the oldest populations. Our contributions are as follows:

- Trying to improve fairness without sacrificing overall performance, we present a novel post-hoc bias mitigation method, Calibration and Group Threshold Optimization (C-GTOP), which combines Platt scaling and group-specific threshold optimization to fix model calibration and achieve improved performance for the oldest age group. Empirical results evidence this behavior.
- We show preliminary empirical evidence of the benefits of C-GTOP over a standard technique like CPP, which severely sacrifices accuracy to mitigate bias.

2   Methods

2.1   Addressing Overdiagnosis

To improve calibration and mitigate overdiagnosis in 3D CNN-based Alzheimer's disease diagnosis classifiers for the elderly, we present Calibration and Group-specific Threshold Optimization (C-GTOP). Our method integrates Platt scaling [14] for initial probability calibration, followed by group-specific threshold optimization to address demographic-specific biases. This combined approach ensures both accurate calibration and effective bias mitigation, tailored to the needs of the elderly population.

Calibration. Let us consider a validation set of n samples $D_{\mathrm{val}} = (X_{\mathrm{val}}, G_{\mathrm{val}}, Y_{\mathrm{val}})$, where X_{val} represents the 3D brain images, G_{val} denotes the protected age groups, and Y_{val} are the binary labels for Alzheimer's diagnosis. Given a 3D CNN classifier h, $p_{\mathrm{val}} = h(X_{\mathrm{val}})$ represents the uncalibrated probabilities, i.e., $p_{i,\mathrm{val}}$ is the probability that individual i ($i \in [1, \ldots, n]$) suffers from AD according to h. Platt scaling, a post-processing calibration method, fits a logistic regression model to the uncalibrated probabilities p_{val} from the validation set and their corresponding true labels Y_{val}. With this model, new probability values $\hat{p}_{\mathrm{val}}$ can be obtained. These probabilities usually show better calibration. This logistic regression model is applied to the test set $D_{\mathrm{test}} = (X_{\mathrm{test}}, G_{\mathrm{test}})$ with uncalibrated probabilities $p_{\mathrm{test}} = h(X_{\mathrm{test}})$, producing the calibrated probabilities $\hat{p}_{\mathrm{test}}$ from p_{test}. After calibration, better confidence estimates can be obtained and model's misclassification reduced. The binarized prediction $\hat{Y} = 1[\hat{p} \geq \tau]$ is then obtained by comparing the calibrated probabilities $\hat{p}$ to a threshold τ. Standard models typically use a single threshold τ for all demographic groups. However, as the distribution of predicted probabilities $\hat{p}$ may differ across demographic groups due to different reasons, applying a different threshold τ_g per group can lead to improved performance [4,8].

Group-Specific Threshold Optimization. To address the issue of biased classification due to varying probability distributions across different age groups, we implement a group-specific threshold optimization strategy. The optimal threshold τ for each age group g is determined by minimizing the difference between the group's false positive rate, $\mathrm{FPR}_g(\tau)$ and the general population's $\mathrm{FPR}_{\mathrm{all}}(0.5)$ at the standard 0.5 threshold, while ensuring that the group's false negative rate $\mathrm{FNR}_g(\tau)$ does not exceed the general population's $\mathrm{FNR}_{\mathrm{all}}(0.5)$. The optimization problem is formulated as follows:

$$\begin{aligned} \underset{\tau}{\text{minimize}} \quad & |\mathrm{FPR}_g(\tau) - \mathrm{FPR}_{\mathrm{all}}(0.5)| \\ \text{subject to} \quad & \mathrm{FNR}_g(\tau) \leq \mathrm{FNR}_{\mathrm{all}}(0.5) \\ & 0.5 \leq \tau \leq 1 \end{aligned}$$

where $\mathrm{FPR}_g(\tau) = 1 - \mathrm{CDF}_{\mathrm{CN},g}(\tau)$ and $\mathrm{FNR}_g(\tau) = \mathrm{CDF}_{\mathrm{AD},g}(\tau)$. Here, $\mathrm{CDF}_{\mathrm{CN},g}(\tau)$ represents the cumulative distribution function of the calibrated probabilities for the CN group at threshold τ, and $\mathrm{CDF}_{\mathrm{AD},g}(\tau)$ represents the

CDF for the AD group, defined as follows:

$$\text{CDF}_{\text{CN},g}(\tau) = P(\hat{p}_{\text{val}} \leq \tau \mid Y_{\text{val}} = \text{CN})$$

$$\text{CDF}_{\text{AD},g}(\tau) = P(\hat{p}_{\text{val}} \leq \tau \mid Y_{\text{val}} = \text{AD})$$

To estimate the distributions, we consider gamma, Student's t, and normal distributions, along with nonparametric methods such as kernel density estimation (KDE) and Gaussian mixture models. We select the distribution that achieves the highest likelihood of fitting the calibrated probabilities $\hat{p}_{\text{val}}$ from the validation set D_{val}. Instead of using a direct optimization solver, we adopt a grid search method, which iteratively evaluates a range of threshold values with τ in the range of $[0.5, 1]$ and selects the one that minimizes the objective function. This method is preferred because it simplifies the optimization process, handling the constraints more flexibly. By limiting τ to the range of $[0.5, 1]$, this strategy aims to prevent an unacceptably high FPR. The optimal τ found is then applied to the calibrated probabilities $\hat{p}_{\text{test}}$ of the test set D_{test} to obtain the final predictions (AD or CN).

Figure 1 shows the histogram of the calibrated probabilities for positive (AD) and negative (CN) samples across different age groups in the validation set, and their CDFs obtained using KDE. The distributions are shown for the entire cohort and for older subgroups (75–84 and 85+ years). A notorious difference in the distribution of predicted probabilities between the age groups exists. When a single classification threshold at $\tau = 0.5$ is applied, the false positive rate (FPR) for both the 75–84 and 85+ age groups is influenced by the CDF of the CN group, such that a lower $\text{CDF}_{\text{CN}}(0.5)$ leads to a higher number of false positives (FP) and a lower number of true negatives (TN) w.r.t. the general population, indicating a higher FPR. Conversely, the optimal threshold for the 85+ group provided by C-GTOP is $\tau_{85+} = 0.68$. The larger $\text{CDF}_{\text{CN}}(\tau_{85+})$ leads to a reduced number of FP and an increased number of TN, thereby reducing the overdiagnosis rate (FPR). However, it does not affect $\text{CDF}_{\text{AD}}(\tau_{85+})$, meaning that the false negative rate (FNR) remains stable. For the 75–84 age group, the optimal threshold provided by C-GTOP is $\tau_{75-84} = 0.65$. Here, a similar behavior is observed: $\text{CDF}_{\text{CN}}(\tau_{75-84})$ increases, while $\text{CDF}_{\text{AD}}(\tau_{75-84})$ increases slightly. This means the FNR increases slightly but remains below the all-ages level. This adjustment maintains an acceptable balance between FPR and FNR, effectively mitigating the overdiagnosis rate (FPR) in elderly populations. Figure 1 highlights the necessity of tailored calibration and threshold optimization for addressing overdiagnosis bias in elderly populations by demonstrating the distinct distributions of predicted probabilities and their impacts on FPRs and FNRs, underscoring the motivation for developing C-GTOP.

2.2   Fairness Definitions and Benchmark

In the context of AI-driven medical diagnostics, fairness is a critical concern. Our method, C-GTOP, adheres to the principle of *minimax fairness* [5], which

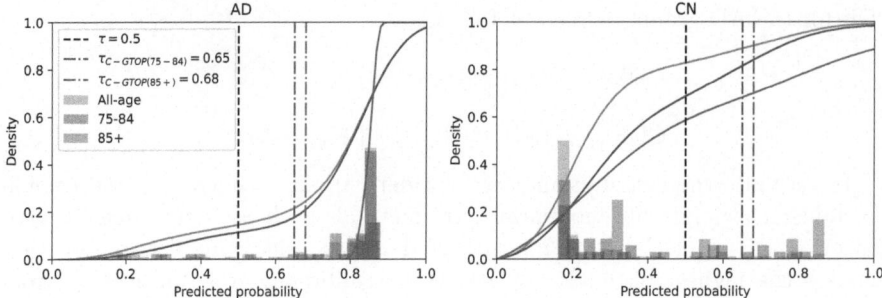

Fig. 1. Histograms and cumulative distribution functions (CDFs) of predicted probabilities for positive samples (AD) and negative samples (CN), segmented by age groups (75–84, 85+). Vertical lines represent different thresholds: $\tau = 0.5$ (black dashed line) and $\tau_{\text{C-GTOP}} = 0.65$ and 0.68 for the 75–84 and 85+ groups respectively (green and red dash-dotted lines). (Color figure online)

seeks to improve the performance of the worst-off group by minimizing the maximum error rate across all groups. This approach is chosen to ensure that the most disadvantaged groups, such as the elderly in our study, receive equitable diagnostic performance. It is important to note that while minimax fairness has been increasingly recognized, *group fairness* [9] remains the most common approach. Group fairness aims to achieve equal performance metrics across protected groups, ensuring that subgroups defined by protected attributes have similar error rates. One popular representative of group fairness is the Calibrated Equalized Odds Post-Processing (CPP) method. This method modifies the model's score outputs for different subgroups to meet an equalized odds objective. However, it is not feasible to satisfy this constraint for both FPR and FNR simultaneously. In our application, we focus on overdiagnosis, aiming to equalize the FPR among age subgroups without reducing diagnostic accuracy for any group, adhering to the group fairness definition. Therefore, we set a cost-constraint objective to achieve equal FPRs among the age subgroups for CPP. Conversely, our minimax fairness approach aims to reduce overdiagnosis bias for the worst-off group, potentially making it more applicable in this clinical setting. To validate this, we compare the results of C-GTOP and CPP.

Many group fairness definitions are incompatible with each other, a concept known as the "impossibility theorem" [2]. Most notably, given an imperfect classifier that outputs a risk score and different base rates between protected groups, it is not possible to achieve calibration within all groups and equalized odds simultaneously in the probabilistic sense [10]. Therefore, we also examine how CPP affects the calibration of each group and compare its performance with C-GTOP in terms of calibration before proceeding to threshold optimization.

3   Experimental Setting

3.1   Data and Protected Groups

We utilize T1-weighted magnetic resonance imaging (MRI) volumes from a combined set of ADNI-1, ADNI-2, and ADNI-GO cohorts within the ADNI dataset [7], comprising 336 and 330 subjects diagnosed with AD and CN, respectively. This research focuses on ensuring fairness concerning the protected attribute of age, categorized into four intervals (< 65, 65–74, 75–84, 85+). Detailed summary statistics for the age distribution of the dataset are presented in Table 1, while statistics for other sensitive attributes, such as sex and ethnicity, can be found in our previous work [3].

Table 1. Age distribution of CN and AD subjects.

	< 65	65 - 74	75 - 84	85+
CN (N%)	11 (3.3%)	163 (49.4%)	117 (35.5%)	39 (11.8%)
AD (N%)	44 (13.1%)	99 (29.5%)	136 (40.5%)	57 (17.0%)

3.2   Metrics

Our study considers overdiagnosis as the main fairness concern. To evaluate model biases in overdiagnosed patients, we compare overdiagnosis rates across different age subgroups within the overall population. Overdiagnosis is defined as the FPR of the binarized model prediction, that is, the probability of indicating a diagnosis when the disease is not actually present. Additionally, we assess the underdiagnosis rate, defined as the FNR to explore the type-2 trade-off between these two fairness aspects following bias mitigation. We also examine the type-1 trade-off between performance and fairness. We use a variety of metrics to evaluate model performance on the test set for each age group. These include threshold-free metrics such as the Area under the ROC curve (AUROC) and Binary Cross-Entropy (BCE), as well as threshold-specific metrics like Balanced Accuracy (BAcc). To assess calibration bias across subgroups, we utilize calibration curves to visually inspect model calibration and quantify it using the Expected Calibration Error (ECE) [11].

3.3   Models

We opt to use the already-built 3D CNN binary model shared by [17], designed to diagnose whether a subject has Alzheimer's disease (AD) or not. This model consists of five convolutional blocks and three fully connected layers. We reserve a 30% of the dataset for testing. The remaining 70% is further divided into 5 folds for cross-validation. The sample assignment into folds follows the procedure in [17]. Five distinct CNN models are generated through this process. Previously, [3] found that these 5 3D CNN-based diagnosis models exhibit poor calibration and a high overdiagnosis rate for older groups, leading to unreliable confidence estimates and overdiagnosis. Therefore, they are an ideal choice to examine the effectiveness of our proposed post-hoc disparity mitigation methods, C-GTOP, and its benchmark, CPP. Thus, we follow the setup of [3].

4   Results and Discussion

4.1   Calibration Effectiveness

Our calibration analysis, as displayed in the top rows of Fig. 2, indicates an improvement with the C-GTOP method across different age groups. Platt scaling in C-GTOP effectively calibrates predictions for all age groups. Notably, for the 75–84 age group, the ECEs of C-GTOP and the baseline (second row, rightmost plot) have non-overlapping confidence intervals, indicating a significant improvement (mean difference: 0.068, paired t-test p-value: < 0.05). The calibration curves of C-GTOP, closer to the diagonal line than those of the baseline, corroborate this trend (first row). C-GTOP maintains consistent AUROC and shows better BCE results across all age groups. Although the BCE for the 65–74 group under C-GTOP increases noticeably (paired t-test p-value < 0.05), it is acceptable because it is still significantly lower than the mean BCE for the Baseline across all age groups (paired t-test p-value < 0.05) and remains the lowest among all age groups. As a reference, we also show CPP's results. This popular mitigation technique shows effectiveness in younger groups but fails to make a substantial impact on older groups. Interestingly, while CPP improves calibration for the younger groups, it creates non-overlapping confidence intervals between the ECE of the 65–74 group and its older counterparts (75–84, 85+), thereby violating the group fairness definition despite enhancing the performance of the worst-case group. Furthermore, CPP reduces AUROC (second row, first plot) and increases BCE (second row, second plot) in younger groups (< 65, 65–74), indicating a type-1 trade-off.

4.2   Overdiagnosis and Underdiagnosis Rates

Figure 2 also displays the overdiagnosis (FPR) and underdiagnosis (FNR) rates in the third row. C-GTOP significantly reduces the overdiagnosis rate in older age groups compared to the baseline. For the 75–84 age group, the FPR (third row, second plot) decreases from 0.383 (95% CI: 0.343–0.443) to 0.278 (95% CI: 0.208–0.314), indicating a non-overlapping confidence interval and a significant mean difference of 0.109 (paired t-test p-value < 0.05). Although the FNR for this group (third row, third plot) slightly increases from 0.142 (95% CI: 0.113–0.170) to 0.172 (95% CI: 0.128–0.216), the overlapping intervals (mean difference: -0.03, paired t-test p-value > 0.05, non-significant statistically) suggest this increase is acceptable. The FNR still overlaps with that of the younger

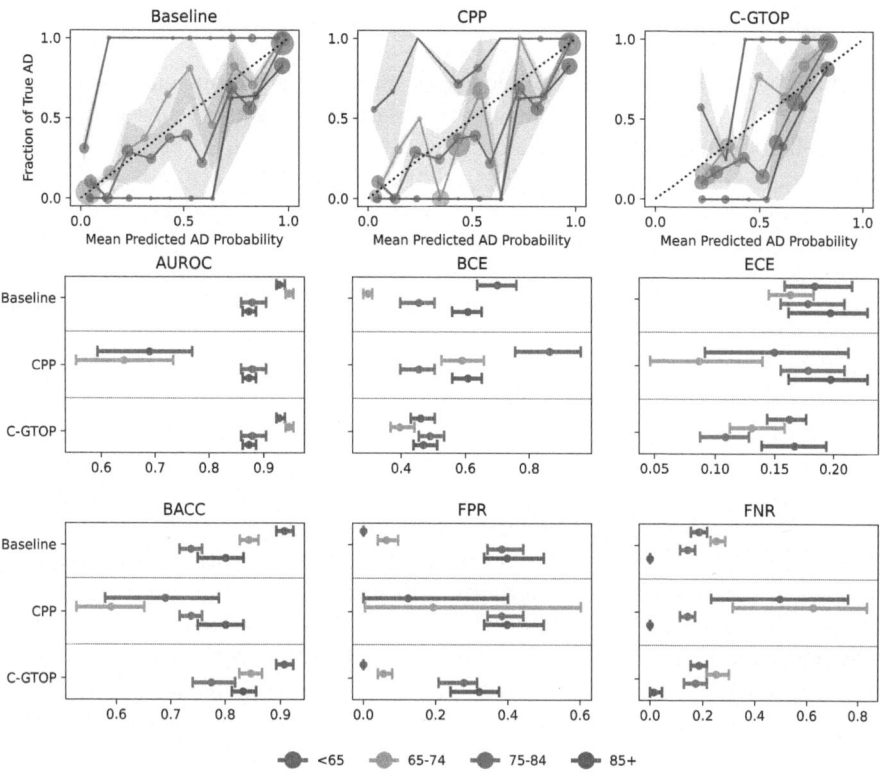

Fig. 2. Comparison of calibration and classification metrics for CNN classifiers: Baseline model (no mitigation) and models after bias mitigation with CPP and C-GTOP. In the calibration curves (first row), point sizes are proportional to the number of samples in each of the 10 bins. In the other rows, error bars represent 95% confidence intervals from 1000 bootstrap iterations.

groups, reflecting a true representation of the underdiagnosis rate of the CNN-based model. For the 85+ age group, the FPR (third row, second plot) reduces from 0.398 (95% CI: 0.333–0.500) to 0.320 (95% CI: 0.241–0.375), with a mean difference of 0.083 (paired t-test p-value > 0.05), showing a noteworthy, though not statistically significant, improvement. The FNR for this group (third row, third plot) experiences a modest increase from 0.0 (95% CI: 0.0 - 0.0) to 0.015 (95% CI: 0.015–0.046). This reduction in FPR from 0.4 to 0.32, while the FNR results in 1.5%, enhances the overall reliability of the model. The detailed metrics, including those broken down by age groups as shown in Fig. 2 and the overall performance metrics of baseline and fairness-enhanced methods (CPP and C-GTOP), are available at https://github.com/ngoc-vien-dang/C-GTOP/tree/main/Results.

4.3   Discussion

Our findings underscore the complexity of achieving simultaneous fairness and calibration in CNN-based Alzheimer's disease diagnosis, as revealed by the "impossibility theorems". We find it necessary to resort to a specifically designed technique, C-GTOP, which combines calibration (via Platt scaling) with group-specific threshold optimization. We provide preliminary evidence that it can effectively improve calibration across all age groups, for example, significantly enhancing ECE for the 75–84 age group. Notably, C-GTOP balances both calibration and fairness. Thus, it is able to reduce the overdiagnosis rate for older groups without harming overall performance, demonstrating its effectiveness in addressing the specific challenges in diagnosing Alzheimer's in elderly populations. As a reference, we also include the results of the popular CPP method. While CPP targets equalized odds, it inadvertently increases calibration gaps between age groups. Specifically, CPP improves calibration for younger groups but creates significant disparities in ECE between the 65–74 group and older groups (75–84, 85+), violating group fairness definitions. Furthermore, CPP increases the overdiagnosis rate for younger groups, resulting in overlapping confidence intervals across age groups, which is not appropriate in a medical context. CPP shows a type-1 trade-off by reducing AUROC and increasing BCE in younger groups (< 65, 65–74), aligning with findings from methods designed to achieve group fairness during training, such as those used in chest X-ray classifiers [18]. For the type-2 trade-off, CPP results in more severe impacts compared to C-GTOP, which demonstrates negligible impacts.

5   Conclusions

C-GTOP is effective in reducing overdiagnosis bias for elderly populations in CNN-based Alzheimer's disease diagnosis. It enhances both fairness and confidence calibration without requiring access to training data. Thus, we expect to further explore its performance in other demographic groups and medical conditions.

Acknowledgements. This work has received funding from the European Union's Horizon 2020 research and innovation programme under Grant Agreement No 952103, EuCanImage.

References

1. Chen, I.Y., Szolovits, P., Ghassemi, M.: Can AI help reduce disparities in general medical and mental health care? AMA J. Ethics **21**(2), 167–179 (2019)
2. Chouldechova, A.: Fair prediction with disparate impact: a study of bias in recidivism prediction instruments. Big Data **5**(2), 153–163 (2017)
3. Dang, V.N., et al.: Auditing unfair biases in CNN-based diagnosis of Alzheimer's disease. In: Workshop on Clinical Image-Based Procedures, pp. 172–182. Springer (2023). https://doi.org/10.1007/978-3-031-45249-9_17

4. Dang, V., Cascarano, A., Mulder, R., et al.: Fairness and bias correction in machine learning for depression prediction across four study populations. Sci. Rep. **14**, 7848 (2024)
5. Diana, E., Gill, W., Kearns, M., Kenthapadi, K., Roth, A.: Minimax group fairness: algorithms and experiments. arXiv preprint arXiv:2011.03108 (March 2021)
6. Ferrante, R.L.M.E.R.: E addressing fairness in artificial intelligence for medical imaging. Nat. Commun. **13**(1), 4581 (2022)
7. Jack, C.R., et al.: The alzheimer's disease neuroimaging initiative (ADNI): MRI methods. J. Magn. Reson. Imaging **27**(4), 685–691 (2008)
8. Jang, T., Shi, P., Wang, X.: Group-aware threshold adaptation for fair classification. In: Proceedings of the AAAI Conference on Artificial Intelligence, vol. 36, pp. 6988–6995 (2022)
9. Kearns, M., Neel, S., Roth, A., Wu, Z.S.: Preventing fairness gerrymandering: auditing and learning for subgroup fairness. In: International Conference on Machine Learning, pp. 2564–2572. PMLR (2018)
10. Kleinberg, J., Mullainathan, S., Raghavan, M.: Inherent trade-offs in the fair determination of risk scores. arXiv preprint arXiv:1609.05807 (November 2016)
11. Nixon, J., Dusenberry, M.W., Zhang, L., Jerfel, G., Tran, D.: Measuring calibration in deep learning. In: CVPR Workshops, vol. 2 (2019)
12. Petersen, E., et al.: Feature robustness and sex differences in medical imaging: a case study in MRI-based alzheimer's disease detection. In: MICCAI 2022: 25th International Conference, pp. 88–98 (2022)
13. Pfohl, S., Marafino, B., Coulet, A., Rodriguez, F., Palaniappan, L., Shah, N.H.: Creating fair models of atherosclerotic cardiovascular disease risk. In: Proceedings of the 2019 AAAI/ACM Conference on AI, Ethics, and Society, pp. 271–278 (2019)
14. Platt, J., et al.: Probabilistic outputs for support vector machines and comparisons to regularized likelihood methods. In: Advances in Large Margin Classifiers, vol. 10, pp. 61–74 (1999)
15. Pleiss, G., Raghavan, M., Wu, F., Kleinberg, J., Weinberger, K.Q.: On fairness and calibration. In: Conference on Neural Information Processing Systems (2017)
16. Qiu, S., et al.: Development and validation of an interpretable deep learning framework for alzheimer's disease classification. Brain **143**(6), 1920–1933 (2020)
17. Wen, J., et al.: Convolutional neural networks for classification of alzheimer's disease: overview and reproducible evaluation. Med. Image Anal. **63**, 101694 (2020)
18. Zhang, H., Dullerud, N., Roth, K., Oakden-Rayner, L., Pfohl, S., Ghassemi, M.: Improving the fairness of chest x-ray classifiers. In: Conference on Health, Inference, and Learning, pp. 204–233. PMLR (2022)

Positive-Sum Fairness: Leveraging Demographic Attributes to Achieve Fair AI Outcomes Without Sacrificing Group Gains

Samia Belhadj$^{(\boxtimes)}$, Sanguk Park, Ambika Seth, Hesham Dar, and Thijs Kooi

Lunit Inc., Seoul, Republic of Korea
{samia.belhadj,tony.superb,ambika.seth,heshamdar,tkooi}@lunit.io

Abstract. Fairness in medical AI is increasingly recognized as a crucial aspect of healthcare delivery. While most of the prior work done on fairness empha-sizes the importance of equal performance, we argue that decreases in fairness can be either harmful or non-harmful, depending on the type of change and how sensitive attributes are used. To this end, we introduce the notion of positive-sum fairness, which states that an increase in performance that results in a larger group disparity is acceptable as long as it does not come at the cost of individual sub-group performance. This allows sensitive attributes correlated with the disease to be used to increase performance without compromising on fairness.

We illustrate this idea by comparing four CNN models that make different use of the race attribute in the training phase. The results show that remov-ing all demographic encodings from the images helps close the gap in perfor-mance between the different subgroups, whereas leveraging the race attribute as a model's input increases the overall performance while widening the disparities between subgroups. These larger gaps are then put in perspective of the collective benefit through our notion of positive-sum fairness to distinguish harmful from non harmful disparities.

Keywords: Fairness · Computer-aided diagnosis · Chest x-ray · Machine Learning

1 Introduction

Medical imaging plays a critical role in diagnosis, treatment planning, and monitoring patient progress. However, the reliability of medical imaging algorithms is not uni-formly distributed across different demographic groups, raising concerns about fairness and potential biases in the results. Fairness in medical imaging most often refers to the equitable treatment of patients from diverse demographic backgrounds, regardless of their gender, race, ethnicity, or other characteristics sensitive to discrimination [19,39].

This equitable treatment is often interpreted as a similar performance across dif-ferent demographic subgroups. When applied to domains like credit card scoring or

---

S. Belhadj and S. Park—These authors contributed equally to this work.

E. Puyol-Antón et al. (Eds.): FAIMI 2024/EPIMI 2024, LNCS 15198, pp. 56–66, 2025.
https://doi.org/10.1007/978-3-031-72787-0_6

AI-powered recruiting, ignoring all sensitive attributes and prioritizing a similar performance across the different demographic subgroups is an acceptable approach. However, in the medical field, demographic attributes are important clinical factors which radiologists and clinicians often take into consideration as they can have a strong impact on their diagnoses and can guide them to consider specific tests or treatments based on the patient's demographic profile. The prevalence of diseases can be correlated to demographic attributes. For example, studies have shown that breast cancer has a higher incidence among Ashkenazi Jewish women [31,38]. And, due to historical and social disparities as well as different physiological features across demographic subgroups, the difficulty level of medical tasks is not uniformly distributed. For this reason, even collecting more or more diverse data does not necessarily produce equal performance across demographic subgroups as the best achievable result is not the same for each of them [28]. In a domain where each improvement can save lives, it is hard to disregard the benefit of the population as a whole for the sake of decreasing the disparities between subgroups.

Petersen et al. [27] examined various types of demographic invariance in medical imaging AI, highlighting why they can be undesirable and stressing the need for better fairness assessments and mitigation techniques in this field. Several fairness measures suffer from degradation in the overall performance by penalizing the performance of an AI system for groups that it performs better on, in order to achieve parity with groups it performs worse on, which is referred to as "levelling down" [25]. While we are aware of papers suggesting training methods which aim to maximize the benefit of each subgroup (Berk Ustun [35], for instance, suggested debiasing methods following the ethical principles of beneficence ("do the best") and non-maleficence ("do not harm") [36] in regards to fairness), and methods which improve fairness by understanding and mitigating the demographic encodings present in images [3,40], we could not find any fairness evaluation framework or definition which allows to compare different models from the prism of harmful and non harmful disparities.

We, therefore, introduce the notion of *positive-sum fairness*: when looking at a situation where we have an initial model and are looking at the trade-off between fairness and performance while trying to improve it, inequitable performance can be acceptable as long as it does not come at the expense of other subgroups and allows a higher overall performance to be achieved. Specifically, we argue that differences in performance can be *harmful* and *non-harmful*. We consider a disparity harmful if it comes at the cost of the overall performance *or* if improving the overall performance is achieved by decreasing performance on any protected subgroup. A difference in performance across protected subgroups is considered non-harmful if, by improving an AI system's performance, we exacerbate the disparities between subgroups without negatively impacting any specific subgroup. This main idea is summarized in Fig. 1.

We compare the positive-sum fairness framework with a more traditional group fairness definition, which is the largest disparity in performance across subgroups. We show that some models, while increasing this disparity, actually improve the performance of each subgroup individually and that other models which decrease the disparity ("improving fairness" from a classic point of view) are harming some subgroups to achieve it.

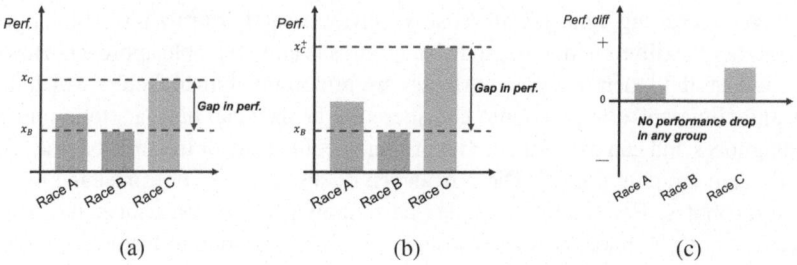

Fig. 1. We investigate fairness of AI models and introduce the concept of 'positive-sum fairness' to differentiate *harmful* and *non-harmful* disparities. Graph a) shows the performance of an initial model per protected groups. b) shows the performance of an updated model with a higher overall performance but a lower fairness, under its standard definition, as indicated by the larger difference between the most and least advantaged groups and therefore could be rejected on the basis of fairness. c) shows the same updated model as b) however it shows the performance difference per group compared to the initial model. In this positive-sum framing we see that none of the groups had a reduction in performance and therefore the increased performance in Race C did not come at the cost of performance in any other group.

2   Related Work

Bias is commonly identified in medical image analysis applications [39,41]. For instance [6], a CNN trained on brain MRI resulted in a significant difference between ethnicities. Seyyed-Kalantari et al. [33] observed that minorities received higher rates of algorithmic underdiagnosis. Zong et al. [41] assessed bias mitigation algorithms in- and out-of-distribution settings. The experiments demonstrated the wide existence of bias in AI-based medical imaging classifiers and none of the bias mitigation algorithms was able to prevent this.

Different definitions of fairness are used:

- **Individual fairness** [26] requires that similar individuals should be treated equally and thus have similar predictions. For example, a model should have comparable diagnosis on two similar X-Ray images.
- **Group fairness** requires equal performance on sub-groups divided based on sensitive attributes (e.g., race, sex, and age). Common group fairness metrics are demographic parity [8], equal odds [12] and predictive rate parity or sufficiency [21].
- **Minimax fairness** [5] seeks to ensure that the worst-off group is treated as fairly as possible, reducing the most severe negative impacts of a decision or system.

These definitions have pros and cons [37]. Individual fairness relies on the choice of the distance metric, which requires expert input. In minimax fairness, the ideal solution is difficult to compute and the degree of unfairness relies heavily on the choice of the set of models. Group fairness metrics are easy to implement and understand, but are not always adapted to the problem nor compatible with one another [2,18]. And even though prior work has broadened the group fairness notion by adding other normative choices than strict equality [1], none of the proposed metrics prevent the harm that could be brought to each subgroup's performance individually or to the whole population's benefit.

As mentioned in the introduction, similarly to [25,27,28,35], we believe that medical AI is different from other domains in that each improvement can save lives. Therefore, increasing disparities to achieve the best performance possible for each demographic subgroup and for the population as a whole could be justified. Previous research has shown that images themselves could carry demographic encodings [9,10]. E.g., Yang et al. [40] investigate the utilization of demographic encodings by analyzing the use of demographic shortcuts for disease classification. Two papers [11,22] examine the relevance of explicitly using sensitive attributes in fair classification systems for non-medical problems. They compare different models which leverage sensitive attributes with a model which is not trained on any sensitive attribute.

3   Methods

3.1   Positive-Sum Fairness

We introduce the principle of positive-sum fairness, which analyzes fairness from the prism of *harmful* and *non harmful* disparities. When looking at changes in model performance and disparities between protected subgroups, there are several explanations for a gap in performance between the most and least advantaged subgroups:

- The most advantaged group's performance improved while others' stayed the same,
- All subgroups' performance improved but one of them increased more than others,
- The most discriminated group's performance decreased while others' stayed the same,
- All subgroups' performance decreased, but one of them decreased more than the others, etc.

The first two would not be considered harmful as they allow to improve the general performance without harming any of the subgroups, thus achieving a collective benefit.

Definition. Positive-sum fairness is a fairness evaluation framework where the goal is to find solutions that increase the overall benefit for all parties together while trying to ensure no one is worse off and ideally, everyone is better off. It looks at the situation where we have an initial model and are looking at the trade-off between fairness and performance when trying to improve the model. Unlike other fairness definitions which aim to minimize the disparity between subgroups or maximize the worst performance among subgroups, positive-sum fairness tries to avoid gains to a group which come *at the expense* of another group while maintaining the overall performance.

Let us assume that we compare N models $\{M\}_{i=1}^{N}$ to a baseline $M_{baseline}$ on K demographic subgroups. And let us consider *measure(M)* as the metric that measures the performance of a model M. Following the positive-sum fairness definition, selecting the best model is equivalent to finding the best trade-off between:

- maximizing the performance gain: $\max_{1 \leq i \leq N} measure(M_i) - measure(M_{baseline})$

– maximizing the smallest performance gain across the subgroups :

$$\max_{1 \leq i \leq N}(\min_{1 \leq k \leq K} measure(M_i)(group_k) - measure(M_{baseline})(group_k))$$

Depending on the task, one could set hard constraints like ensuring there is no performance loss for any subgroup (aka the selected model M_i should ensure that $\min_{1 \leq k \leq K} measure(M_i)(group_k) - measure(M_{baseline})(group_k) \geq 0$) and the overall performance is improved (aka the selected model M_i should ensure that $measure(M_i) - measure(M_{baseline}) \geq 0$) or find the most relevant trade-off between the two optimization problems.

3.2   Application

To put this fairness notion into practice and show the difference with traditional group fairness, we compare three models which use sensitive attributes to a baseline model. The way sensitive attributes are used by the model is known to have an impact on the fairness and performance of the model [3,11,22,40]. Therefore, we make use of models that explicitly include sensitive attributes, or conversely, remove any demographic encoding from the input data.

The four models are trained on a multi-label classification problem of findings in chest radiography (CXR). In all settings, a Densenet-121 [13] backbone is used, which was empirically determined to give the best performance for this problem. The exact model architectures are shown in Fig. 2 and described below:

– **M1**: a baseline classifier using the images as input and trained to predict the targeted CXR findings associated to our dataset. The model comprises a backbone to extract the image features and a finding branch consisting of a fully connected layer and a binary cross entropy loss for each finding.
– **M2**: a classifier using both the images and race features as input. The race information comes in the form of a categorical variable, which we convert to a one-hot vector and feed to a fully-connected layer. We concatenate the features from the fully connected layer and the image features before forwarding to finding branch. The model is trained end-to-end.
– **M3**: a classifier using the images as input only, but trained to predict image findings as well as the race group (i.e. this model aims to exploit the race encodings present in the images). For this model, we modify the final layer of the baseline classifier by adapting the loss function to optimize the two tasks: CXR findings and race group. We also transform race information to one-hot encoded vector to apply multi-class loss. The race classification branch is made of a fully-connected layer and a cross entropy loss function. The final loss is calculated by adding finding loss and the race loss with a loss weight λ.

$$L(y_{cxr}, y_{race}) = L(y_{cxr}) + \lambda L(y_{race})$$

– **M4**: a classifier using the images as input, trained to predict image findings, while minimizing the use of race information encoded in the image. For this model, we implement the gradient reversal technique described in [29]. We apply the gradient reversal layer before the race branch.

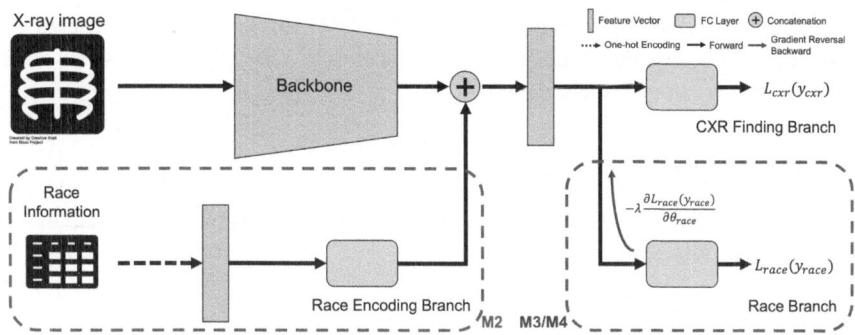

Fig. 2. To investigate the effect of sensitive attributes on performance and fairness, we evaluate four different model architectures, denoted M1, M2, M3 and M4. M1, the baseline, has a backbone and classification. M2 has a race encoding branch to learn race-encoded features directly from metadata. M3 and M4 have an additional race branch to predict the race group which is implicitly encoded in the image, from the image features. The difference between M3 and M4 is that we add a gradient reversal layer before the race branch.

4 Experiments

Data. We use chest radiographs from MIMIC-CXR-JPG [16,30]. The dataset has annotations for 14 findings. However, we focus on lung lesions, pneumonia, pleural effusion and consolidation as the diseases associated with these findings have been shown to be correlated with ethnicity [4,17,34]. We use only frontal images and split the dataset into training, validation, and test sets on a patient level. In total, 237,972, 1,959, and 3,403 images are used for training, validation, and testing, respectively.

Sensitive Attributes. We define the protected subgroups based on the self-reported race from MIMIC-IV [14,15] and split it into five groups: White, African-American, Latino, Asian, others.

Model Training. We train our 4 models to predict all 14 CXR findings and a race group. We initialize a DenseNet-121 backbone with pre-trained weights from ImageNet [32]. The images are resized to 256×256, and augmented using random rotation from [-15,15] degree range and random horizontal flip. We conduct the experiments with 8 V100 NVIDIA GPU. AdamW [24] is used with an initial learning rate of 0.002 which is updated using the cosine annealing warm up [23] scheduler.

Evaluation. We compare the four models by general performance and fairness across the protected subgroups. The general performance is assessed using the Area under the ROC curve (AUROC) score and the traditional group fairness metric used to compare with positive-sum fairness is expressed by (1 - largest disparity between protected subgroups in terms of AUROC) [20]. We use the AUROC mean and confidence bounds generated using bootstrapping with 300 samples [7]. We do not consider protected subgroups which have less than 5 positive cases or less than 5 negative cases as this results in poor estimates of performance.

4.1 Initial Results

According to traditional group fairness, in assessing the results of the four models shown in Fig. 3a one could conclude that:

M2 Improves the Overall Performance. Our results show that M2 outperforms M1 in terms of AUROC. This is in line with our expectation as we are providing an additional relevant medical feature for the model to learn from. This better performance comes with a larger gap in AUROC between the most advantaged and most discriminated races, in other words less fairness from a traditional point of view. But this larger disparity is not necessarily *harmful* according to the positive-sum fairness as we will discuss it in the next section.

M4 Improves the Fairness. M4 improves fairness for lung lesions and consolidations, while performing similar for pneumonia and pleural effusion. The improved fairness is likely due to the gradient reversal layer, which removes race information from the image and prevents the model from exploiting any demographic shortcut.

No Clear Pattern for M3. The results for M3 are less consistent. Its performance is lower than the baseline except for pneumonia and its fairness measurement is sometimes higher and other times lower than the baseline's. If the baseline model exploited demographic encodings present in the images to generate shortcuts, training M3 to maximize the race prediction might have intensified the impact of these shortcuts.

4.2 Positive-Sum Fairness

We now apply the notion of positive-sum fairness, defined in Sect. 3.1 and reframe the fairness vs performance problem as shown in 3b. Here, the x-axis represents the difference in performance between each improved classifier and the baseline ($AUROC(M_i) - AUROC(M_1)$) and the y-axis shows the performance increase (or decrease) for the least improved subgroup ($\min_{1 \leq k \leq K} AUROC(M_i)(race_k) - AUROC(M_1)(race_k)$). A negative value indicates that the model performs worse for the given subgroup.

Any classifier which has the exact same overall performance and exact same performance per protected subgroup (race) as the baseline, would be at coordinate (0,0). Any classifier that has a negative x-coordinate, would have a lower general performance than M1 and any classifier that has a negative y-coordinate would have at least one protected subgroup with a lower AUROC than M1 (at least one subgroup negatively impacted by the changes brought to the baseline model).

For lung lesions, Fig. 3b shows that M2 appears in the positive side of the x and y axes, meaning that the performance was improved without harming any subgroup's

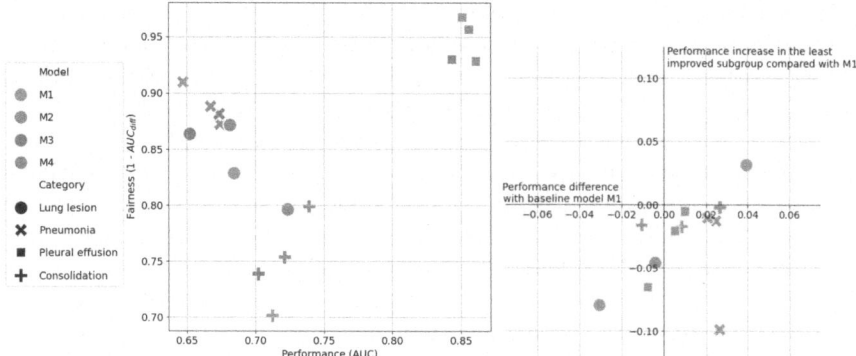

(a) Traditional group fairness vs performance frame-   (b) Positive-sum fairness framework
work

Fig. 3. We put in parallel 2 different fairness vs performance frameworks: in figure (a), we compute both the performance (AUROC) and fairness (as 1 - the difference in AUROC between the most and least advantaged groups) of the 4 models per lesion. And in figure (b), we show, the difference in overall performance and in performance per protected subgroup between the 3 improved classifiers and the baseline M1. The x axis compares the performance of each improved classifier with the baseline and the y axis shows whether at least one protected subgroup has been harmed by the modifications brought to the baseline classifier.

performance. And this even though the Fig. 3a shows a decrease in fairness (larger disparity between the most advantaged and least advantaged subgroups) for M2 compared with M1. This matches the previous conclusion that the larger performance gap between protected subgroups for M2 compared with M1 cannot be considered harmful as every protected subgroup's performance was individually increased.

On the other hand, for lung lesions, model M4 improved fairness (smaller disparity between the most advantaged and least advantaged subgroups) as shown in the Fig. 3a. However, the Fig. 3b, shows that M4 has negative y coordinates, meaning that at least one subgroup was harmed while trying to achieve a smaller disparity between protected subgroups.

5   Conclusion

In this paper, we presented the notion of *positive-sum fairness* and argued that larger disparities are not necessarily harmful, as long as it does not come at the expense of a specific subgroup performance. The general performance, standard fairness and positive-sum fairness of four models was analyzed, each leveraging sensitive attributes in a different way.

Our study highlights the need for a nuanced understanding of fairness metrics and their implications in real-world applications. Good incorporation of medical knowledge is crucial when utilizing sensitive information and evaluating fairness accurately, particularly in cases where models may show a large performance disparity.

When traditional methods often aim for equality, positive-sum fairness focuses on equity, pushing for each group to achieve its highest possible performance level. This can lead to better overall outcomes, as it encourages to address the specific needs and challenges of each group without diminishing the quality of care for others. But, being defined as an optimization problem, it could also have unintended side effects as it may inadvertently prioritize larger or more well-represented groups by focusing the efforts on the groups with the highest impact on the overall performance rather than those with the most critical needs. Therefore, it is to be noted that meeting the positive-sum fairness criterion alone does not ensure a model to be fair from an egalitarian perspective, and the use of this notion in conjunction with other metrics can give a more holistic understanding of a model's fairness.

As positive-sum fairness is a relative measure, it requires a baseline to be used. Further work in this area would include developing a more robust baseline or adapting the approach to remove the need for a baseline. It would also be worth it to compare out-of-domain tested models, include other sensitive attributes such as sex and age and take into account confounding factors.

Disclosure of Interests. The authors declare that there are no conflicts of interest regarding the publication of this paper.

References

1. Baumann, J., Hertweck, C., Loi, M., Heitz, C.: Distributive justice as the foundational premise of fair ML: unification, extension, and interpretation of group fairness metrics. arXiv:2206.02897 (2023)
2. Berk, R., Heidari, H., Jabbari, S., Kearns, M., Roth, A.: Fairness in criminal justice risk assessments: the state of the art. arXiv:1703.09207 (2017)
3. Brown, A., Tomasev, N., Freyberg, J., Liu, Y., Karthikesalingam, A., Schrouff, J.: Detecting shortcut learning for fair medical AI using shortcut testing. arXiv:2207.10384 (2022)
4. Burton, D.C., et al.: Socioeconomic and racial/ethnic disparities in the incidence of bacteremic pneumonia among US adults. Am. J. Public Health **100**(10), 1904–1911 (2010)
5. Diana, E., Gill, W., Kearns, M., Kenthapadi, K., Roth, A.: Minimax group fairness: algorithms and experiments. arXiv:2011.03108 (2021)
6. Stanley, E.A.M., Wilms, M., Mouches, P., Forkert, N.D.: Fairness-related performance and explainability effects in deep learning models for brain image analysis. J. Med. Imaging **9**(6), 061102 (2022)
7. Efron, B.: Better bootstrap confidence intervals. J. Am. Stat. Assoc. **82**(397), 171–185 (1987)
8. Feldman, M., Friedler, S., Moeller, J., Scheidegger, C., Venkatasubramanian, S.: Certifying and removing disparate impact. arXiv:1412.3756 (2015)
9. Gichoya, J.W., et al.: AI recognition of patient race in medical imaging: a modelling study. Lancet Digit. Health **4**(6), e406–e414 (2022)
10. Glocker, B., Jones, C., Bernhardt, M., Winzeck, S.: Algorithmic encoding of protected characteristics in chest X-ray disease detection models. EBioMedicine **89**(104467), 104467 (2023)
11. Haeri, M.A., Zweig, K.A.: The crucial role of sensitive attributes in fair classification. In: 2020 IEEE Symposium Series on Computational Intelligence (SSCI), pp. 2993–3002 (2020). https://doi.org/10.1109/SSCI47803.2020.9308585

12. Hardt, M., Price, E., Srebro, N.: Equality of opportunity in supervised learning. arXiv:1610.02413 (2016)
13. Huang, G., Liu, Z., van der Maaten, L., Weinberger, K.Q.: Densely connected convolutional networks. arXiv:1608.06993 (2018)
14. Johnson, A., Bulgarelli, L., Pollard, T., Horng, S., Celi, L.A., Mark, R.: MIMIC-IV (2023)
15. Johnson, A.E.W., et al.: MIMIC-IV, a freely accessible electronic health record dataset. Sci. Data **10**(1), 1 (2023)
16. Johnson, A.E.W., et al.: MIMIC-CXR-JPG, a large publicly available database of labeled chest radiographs. arXiv:1901.07042 (2019)
17. Joseph, N.P., et al.: Racial and ethnic disparities in disease severity on admission chest radiographs among patients admitted with confirmed coronavirus disease 2019: a retrospective cohort study. Radiology **297**(3), E303–E312 (2020)
18. Kleinberg, J., Mullainathan, S., Raghavan, M.: Inherent trade-offs in the fair determination of risk scores. arXiv:1609.05807 (2016)
19. Lara, M.A.R., Echeveste, R., Ferrante, E.: Addressing fairness in artificial intelligence for medical imaging. Nat. Commun. **13**, 4581 (2022)
20. Lee, J., Brooks, C., Yu, R., Kizilcec, R.: Fairness hub technical briefs: AUC gap. arXiv:2309.12371 (2023)
21. Lee, J.K., et al.: Fair selective classification via sufficiency. In: International Conference on Machine Learning (2021). https://api.semanticscholar.org/CorpusID:235826429
22. Žliobaitė, I., Custers, B.: Using sensitive personal data may be necessary for avoiding discrimination in data-driven decision models. Artif. Intell. Law **24**(2), 183–201 (2016). https://doi.org/10.1007/s10506-016-9182-5
23. Loshchilov, I., Hutter, F.: SGDR: stochastic gradient descent with warm restarts. arXiv:1608.03983 (2017)
24. Loshchilov, I., Hutter, F.: Decoupled weight decay regularization. arXiv:1711.05101 (2019)
25. Mittelstadt, B., Wachter, S., Russell, C.: The unfairness of fair machine learning: levelling down and strict egalitarianism by default. arXiv:2302.02404 (2023)
26. Mukherjee, D., Yurochkin, M., Banerjee, M., Sun, Y.: Two simple ways to learn individual fairness metrics from data. arXiv:2006.11439 (2020)
27. Petersen, E., Ferrante, E., Ganz, M., Feragen, A.: Are demographically invariant models and representations in medical imaging fair? arXiv:2305.01397 (2024)
28. Petersen, E., Holm, S., Ganz, M., Feragen, A.: The path toward equal performance in medical machine learning. Patterns **4**(7), 100790 (2023). https://doi.org/10.1016/j.patter.2023.100790
29. Raff, E., Sylvester, J.: Gradient reversal against discrimination. arXiv:1807.00392 (2018)
30. Rajeev, C., Natarajan, K.: Data augmentation in classifying chest radiograph images (CXR) using DCGAN-CNN. In: Solanki, A., Naved, M. (eds.) GANs for Data Augmentation in Healthcare. Springer, Cham, pp. 91–110 (2023). https://doi.org/10.1007/978-3-031-43205-7_6
31. Rubinstein, W.S.: Hereditary breast cancer in jews. Fam. Cancer **3**(3–4), 249–257 (2004)
32. Russakovsky, O., et al.: ImageNet large scale visual recognition challenge. arXiv:1409.0575 (2015)
33. Seyyed-Kalantari, L., Zhang, H., McDermott, M., Chen, I., Ghassemi, M.: Underdiagnosis bias of artificial intelligence algorithms applied to chest radiographs in under-served patient populations. Nat. Med. **27**, 2176–2182 (2021). https://doi.org/10.1038/s41591-021-01595-0
34. Shi, H., et al.: Genomic landscape of lung adenocarcinomas in different races. Front. Oncol. **12**, 946625 (2022)
35. Ustun, B., Liu, Y., Parkes, D.: Fairness without harm: decoupled classifiers with preference guarantees. In: Chaudhuri, K., Salakhutdinov, R. (eds.) Proceedings of the 36th International

Conference on Machine Learning. Proceedings of Machine Learning Research, vol. 97, pp. 6373–6382. PMLR (2019). https://proceedings.mlr.press/v97/ustun19a.html

36. Varkey, B.: Principles of clinical ethics and their application to practice. Med. Princ. Pract. **30**(1), 17–28 (2021)

37. Verma, S., Rubin, J.S.: Fairness definitions explained. In: 2018 IEEE/ACM International Workshop on Software Fairness (FairWare), pp. 1–7 (2018). https://api.semanticscholar.org/CorpusID:49561627

38. Warner, E., et al.: Prevalence and penetrance of BRCA1 and BRCA2 gene mutations in unselected ashkenazi jewish women with breast cancer. J. Natl. Cancer Inst. **91**(14), 1241–1247 (1999)

39. Xu, Z., Li, J., Yao, Q., Li, H., Zhou, S.K.: Fairness in medical image analysis and healthcare: a literature survey. TechRxiv (2023). https://doi.org/10.36227/techrxiv.24324979.v1

40. Yang, Y., Zhang, H., Gichoya, J.W., Katabi, D., Ghassemi, M.: The limits of fair medical imaging AI in the wild. arXiv:2312.10083 (2023)

41. Zong, Y., Yang, Y., Hospedales, T.: MEDFAIR: benchmarking fairness for medical imaging. arXiv:2210.01725 (2023)

All You Need Is a Guiding Hand: Mitigating Shortcut Bias in Deep Learning Models for Medical Imaging

Christopher Boland[1,2(✉)], Owen Anderson[1], Keith A. Goatman[1],
John Hipwell[1], Sotirios A. Tsaftaris[1,2], and Sonia Dahdouh[1]

[1] Canon Medical Research Europe, Edinburgh EH6 5NP, UK
christopher.boland@mre.medical.canon
[2] School of Engineering, The University of Edinburgh, Edinburgh EH9 3FG, UK

Abstract. Deep learning models for medical imaging are prone to learning shortcut solutions that rely on spurious correlations instead of clinically meaningful features, leading to poor generalization to new data. We propose an oracle-guided training scheme that encourages a student model to learn robust features in the presence of shortcuts. Our method regulates prediction confidence across intermediate network layers, significantly reducing shortcut impact. We evaluate our approach on CIFAR10, CheXpert, and ISIC 2017 datasets using ResNet18 and DenseNet121 architectures. Consistently, we outperform a model trained using Empirical Risk Minimization on a dataset containing a shortcut. In several cases, we close the gap on our clean baseline to the point that there is no statistically significant difference in performance. We also address the practical challenge of obtaining a clean oracle model, enhancing the method's real-world applicability.

Keywords: Shortcut learning · Bias Mitigation · Medical Imaging · Fairness

1 Introduction

Despite the success of deep learning in medical imaging, concerns about the robustness and safety of AI systems limit their deployment in real-world healthcare settings [13,14,26]. A key challenge is "shortcut learning", where models learn to rely on spurious correlations or biases in their training data instead of meaningful medical features [1,10,20]. In medical imaging, such shortcuts can arise from various sources, such as image acquisition differences, patient demographics, or the presence of medical devices [6,11,21,27]. Models that exploit these shortcuts may achieve high performance on a narrow training distribution but fail to generalize to diverse, real-world clinical settings, potentially leading to unreliable or unsafe predictions.

Empirical Risk Minimization (ERM) is a widely used principle in machine learning, aiming to minimize the average loss over the training data. The simplicity bias of neural networks means ERM can inadvertently lead to models

E. Puyol-Antón et al. (Eds.): FAIMI 2024/EPIMI 2024, LNCS 15198, pp. 67–77, 2025.
https://doi.org/10.1007/978-3-031-72787-0_7

exploiting easy-to-learn dataset biases to minimize loss. Many state-of-the-art techniques mitigate this either by assuming access to explicitly labeled shortcut sources [16,22], which is often infeasible in medical imaging due to the multitude of potential biases, or by only targeting specific kinds of shortcuts, such as small objects or simple features [9,19].

In this paper, we introduce a novel oracle-guided training scheme, where the training of one model (the student) is conditioned using information from a carefully curated oracle model. This approach guides the student towards more robust and generalizable solutions [7,28]. We mitigate shortcut learning by leveraging an oracle trained on clean, unbiased data, thus eliminating the need for explicit shortcut source labeling. The oracle guides the learning process of a student trained on potentially biased data. Additionally, we address the practical challenges of obtaining an oracle and propose a solution, enhancing the method's applicability in real-world clinical settings. We demonstrate the effectiveness of our method on multiple datasets, model architectures, and shortcut types.

2   Background

2.1   Shortcut Learning in Healthcare AI

As interest in shortcut learning has grown, numerous studies have exposed the presence of shortcuts in medical image datasets and highlighted the degradation of AI systems that this prevalence introduces. Medical equipment, such as color calibration charts and rulers in skin lesion images or chest drains visible in X-rays, can be correlated with the class label, causing models to rely on their presence or absence for predictions [11,21] Even subtle features, such as differences in the image acquisition process or patient demographic groups, can introduce shortcuts [2,6].

2.2   Mitigating Shortcut Learning

Existing mitigation approaches can be categorized based on their requirement for explicit shortcut source labels. Methods such as GroupDRO [22] and REPAIR [16] use group annotations to re-weight or resample training data but are limited by the need for shortcut labels. Other works mitigate shortcuts without explicit labels by making assumptions about their nature or impact on model behavior, such as focusing on small, localized features [19] or features detectable by low-capacity models [9].

MIRO is the most closely related method to ours [7]. It uses a pre-trained model as an oracle to improve robustness to domain shifts by maximizing mutual information (MI) between feature vectors extracted from the oracle and student models.

However, MIRO's applicability to shortcut learning in the medical imaging domain is limited. Large, diverse pre-trained models are scarce in medical imaging. While datasets like RadImageNet exist for radiological images, they don't cover the full spectrum of medical imaging modalities [18]. Significant visual differences between medical imaging domains mean that applying an oracle-guided approach would likely require tailoring the oracle to the specific imaging domain. For example, visual features relevant to radiological images may not apply to dermatological or ophthalmological images. We address this by training the oracle on a curated, task-specific dataset devoid of shortcuts. Rather than maximizing MI between feature vectors, which could lead to overfitting, especially with our oracle trained on a smaller dataset, we guide the student model by comparing its ability to distinguish between classes at its intermediary layers to the oracle. This encourages learning features that facilitate similar class inference abilities at each layer, discouraging reliance on overly simple features that allow the model to distinguish between classes too easily.

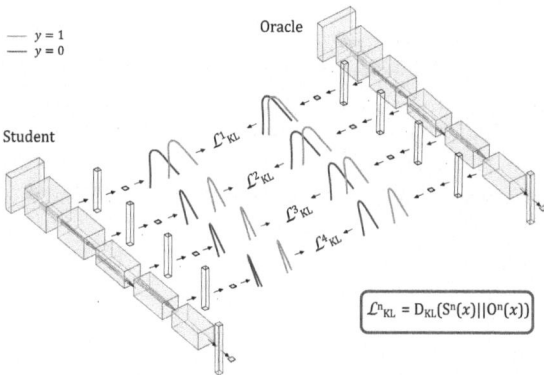

Fig. 1. Overview of the proposed oracle-guided training method. The oracle model, trained on clean data, guides the student model's learning process through comparison of intermediate layer predictions.

3 Method

3.1 Model Confidence

Models trained on biased data tend to exhibit overconfidence in their predictions [24]. This aligns with the nature of shortcut learning: models that exploit shortcuts in their training data do so because they provide an easier path to distinguishing between classes. These simple features allow the model to achieve high confidence with less effort [3]. We are interested in (a) how a model's confidence in inferring class changes through its internal layers and (b) the effect of including shortcuts in its training data. To understand this, we introduce classification probes - linear classifiers attached to the network's intermediate layers. After the network finishes training, these probes are fine-tuned on the downstream task. They serve to provide insight into models' ability to infer the true class at different depths of the network, allowing us to examine how the presence of shortcuts influences class inference capabilities throughout the entire model.

To quantify a model's confidence over a batch at each layer, we follow prior work and consider the output logits of the classification probes [23]. The sigmoid of the output logits ranges from 0 to 1, indicating the likelihood that the input belongs to the positive (1) or negative class (0). An output of 0.5 signifies maximal uncertainty. We define model confidence as the absolute difference between its prediction and 0.5, shown in Eq. 1, where $f(x)$ represents the sigmoid output of the model for input x.

$$C(x) = |f(x) - 0.5| \tag{1}$$

3.2   Oracle-Guided Shortcut Mitigation

Our oracle-guided training scheme (Fig. 1) aims to mitigate shortcut learning by preventing the student model from becoming overconfident in shortcut features. The student model is trained to minimize the empirical loss on a biased dataset while matching the oracle's confidence in inferring the true class at each layer. We encourage this learning behaviour by minimizing the Kullback-Leibler (KL) divergence between the output probability distributions of each model's classification probes.

The training loss of the student is described in Eq. 2, where $\mathcal{L}_{total}$ is the total loss, $\mathcal{L}_{BCE}$ is the Binary Cross Entropy (BCE) loss, $\mathcal{L}_{KL}$ is the total KL divergence loss between the oracle and student probes, and λ is a weight applied to $\mathcal{L}_{KL}$ whose value is determined via a hyperparameter sweep.

$$\mathcal{L}_{total} = \mathcal{L}_{BCE} + \lambda \cdot \mathcal{L}_{KL} \tag{2}$$

Each epoch, after the student model's parameters have been updated, the probes are fine-tuned on the downstream classification task. This maintains the probes' ability to classify based on the learned encodings of the student. Separate optimizers are used for the encoder and probe parameters to prevent unintended interactions between their learning objectives.

3.3   Datasets and Model Architectures

We evaluate our approach on three datasets: CIFAR10, CheXpert, and ISIC 2017. CIFAR10, a common computer vision benchmark, contains $60,000$ 32×32 color images across 10 classes. We use a binary subset of $10,000$ training images for cat vs. dog classification [12]. CheXpert is a large-scale chest radiography dataset. In this work, we consider the task of classifying the presence of a pneumothorax. Our final training dataset consists of $2,914$ training images [11]. Finally, ISIC 2017 is used for skin lesion analysis and presents another medical context for our evaluation. It contains $1,120$ training dermoscopic images for the binary classification of melanoma/seborrheic keratosis versus benign lesions [8].

We consider ResNet18 and DenseNet121 to demonstrate that our method is not architecture-specific. We train models using Stochastic Gradient Descent (SGD) and AdamW optimizers with learning rates of $1 \times e^{-2}$, $1 \times e^{-3}$, and $1 \times e^{-4}$, as the choice of optimizer and learning rate can impact shortcut learning [5]. For CheXpert and ISIC experiments, images are re-sized to ImageNet resolution, 224×224. Each model is trained for 300 epochs with early stopping based on the validation loss to determine the best model. We use 5-fold cross-validation, with consistent test sets across folds. Our experimental setup utilizes Python and Pytorch and we train on an NVIDIA RTX 2080 Ti and a Tesla V100.

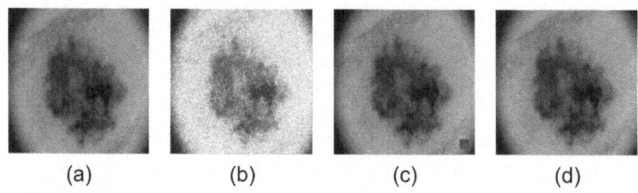

(a)   (b)   (c)   (d)

Fig. 2. ISIC skin lesion image augmented with synthetic shortcuts: (a) original; (b) noise; (c) red square (constant location); (d) red square (random location). (Color figure online)

3.4 Artificial Shortcuts

To test the robustness of our approach against various types of shortcuts, we generate augmented datasets containing two types of synthetic shortcut (Fig. 2):

1. **Diffuse shortcuts:** leveraging random, uniform noise patterns as a spurious signal spread throughout the image. The noise is generated using a uniform distribution with values between 0 and 0.15 applied to each pixel. The noise pattern is generated before training and applied to all shortcutting samples.
2. **Localized shortcuts:** introducing small, red square shapes to the image, similar to other work [9]. On our monochrome dataset (CheXpert), this presents as a white square. We test two variants:
   (a) **Constant location:** the square appears in a fixed spot.
   (b) **Random location:** the square's location varies among images.

In our training splits the shortcuts is perfectly correlated with one class, while it is distributed evenly between both classes in the test and validation splits.

4 Results

In our experiments, we first investigate the impact of shortcut learning on a model's predictive behavior. Next, we evaluate the effectiveness of our oracle-guided training scheme in mitigating shortcut learning. Finally, we explore strategies to reduce the technical burden of obtaining a clean oracle model.

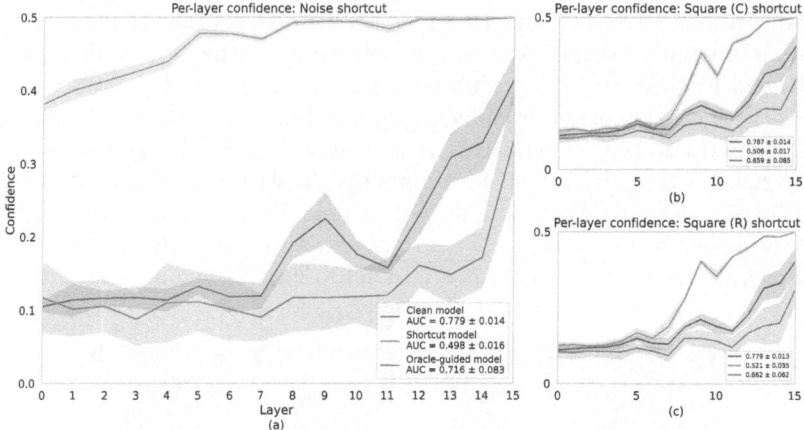

Fig. 3. Confidence per-layer of a ResNet18 trained on the ISIC classification task. We compare a clean model to a model trained in the presence of each shortcut: (a) Noise; (b) constant square; (c) random square. All models are trained with the AdamW optimizer and a learning rate of $1 \times e^{-3}$.

4.1 Shortcuts Cause Overconfidence

Considering the complexity of many medical image analysis tasks, we expect a well-trained model using clinically relevant features to exhibit lower confidence than a model relying on easy shortcut features. Additionally, we might expect that the shortcutting model's confidence will increase in earlier layers, aligning with the expectation that the deeper layers of the network capture more sophisticated features [4]. To test this, we train two ResNet18 models on ISIC using ERM. One is trained on a clean dataset, while the other is trained on a dataset featuring an artificial shortcut associated with the malignant class. Both are evaluated on a "clean" test set where the shortcut is evenly distributed between the classes. After training, we fine-tune our classification probes for each model.

Figure 3 illustrates the per-layer confidence (Eq. 1) of each model. In line with our hypothesis, the shortcutting model becomes overconfident in earlier layers compared to the baseline. We also notice a significant difference in the influence of a diffuse shortcut (noise) from localized shortcuts. In the diffuse case, confidence is very high even from the first layer. This is likely because the diffuse shortcut is comprised of low-level features that require very little disambiguation by the network, unlike the localized shortcuts. We also find training with our proposed method prevents such overconfidence compared to the baseline.

4.2 Oracle-Guided Mitigation of Shortcut Learning

Table 1 compares the performance of our oracle-guided training scheme against ERM trained on shortcut data and a clean benchmark model for a ResNet18. Our method consistently outperforms ERM and closes the gap to the clean baseline across all datasets, learning rates, and optimizers. Additionally, the performance of an ERM model trained on the shortcut dataset is always statistically significantly worse than the clean benchmark ($p < 0.05$), as measured by a two-sample t-test. In several cases, a model trained with our method has no statistically significant difference in performance compared to the benchmark. Furthermore, our method is robust across datasets and optimizers, with consistent improvements over ERM. We observe similar trends with our DenseNet121 model.

Table 1. AUC of a ResNet18. All models are trained with a learning rate of $1 \times e^{-4}$. Samples with no statistically significant difference in performance from the clean benchmark ($p \geq 0.05$) are marked with an asterisk (*). The best results are in bold, and the second best are italicized.

Dataset	Shortcut	Optimizer	ERM (clean)	ERM (shortcut)	Ours
CIFAR10	Noise	SGD	*0.804 ± 0.000*	0.593 ± 0.020	**0.820 ± 0.016***
CIFAR10	Square (C)	SGD	**0.814 ± 0.003**	0.720 ± 0.002	*0.792 ± 0.011*
CIFAR10	Square (R)	SGD	**0.792 ± 0.002**	0.690 ± 0.001	*0.792 ± 0.015 *￼
CIFAR10	Noise	AdamW	**0.837 ± 0.003**	0.661 ± 0.008	*0.815 ± 0.016*
CIFAR10	Square (C)	AdamW	**0.841 ± 0.001**	0.582 ± 0.004	*0.757 ± 0.034*
CIFAR10	Square (R)	AdamW	**0.824 ± 0.004**	0.699 ± 0.008	*0.779 ± 0.017*
ISIC	Noise	SGD	**0.743 ± 0.003**	0.536 ± 0.030	*0.702 ± 0.076*￼
ISIC	Square (C)	SGD	**0.796 ± 0.010**	0.616 ± 0.010	*0.637 ± 0.047*
ISIC	Square (R)	SGD	**0.795 ± 0.009**	0.603 ± 0.021	*0.641 ± 0.086*
ISIC	Noise	AdamW	**0.745 ± 0.014**	0.523 ± 0.027	*0.562 ± 0.014*
ISIC	Square (C)	AdamW	**0.836 ± 0.006**	0.580 ± 0.007	*0.662 ± 0.009*
ISIC	Square (R)	AdamW	**0.829 ± 0.003**	0.584 ± 0.017	*0.676 ± 0.008*
CheXpert	Noise	SGD	**0.643 ± 0.022**	*0.526 ± 0.022*	0.522 ± 0.035
CheXpert	Square (C)	SGD	**0.740 ± 0.007**	0.613 ± 0.012	*0.716 ± 0.015*
CheXpert	Square (R)	SGD	**0.737 ± 0.008**	0.636 ± 0.006	*0.733 ± 0.010*￼
CheXpert	Noise	AdamW	**0.660 ± 0.026**	0.497 ± 0.020	*0.501 ± 0.017*
CheXpert	Square (C)	AdamW	**0.755 ± 0.010**	0.539 ± 0.022	*0.686 ± 0.058*￼
CheXpert	Square (R)	AdamW	**0.755 ± 0.013**	0.607 ± 0.017	*0.744 ± 0.017*￼

4.3   Training the Oracle with Less

Collecting large, shortcut-free datasets for medical applications is challenging and costly. Given the impracticality of guaranteeing a completely clean dataset for oracle training, especially with large datasets, we investigate our method's effectiveness with a non-optimal oracle trained on small, curated datasets. We evaluated oracles trained on 5%, 10%, 20%, and 30% of the full training sets. At the smallest subset (5%), the ISIC dataset has only 56 training images.

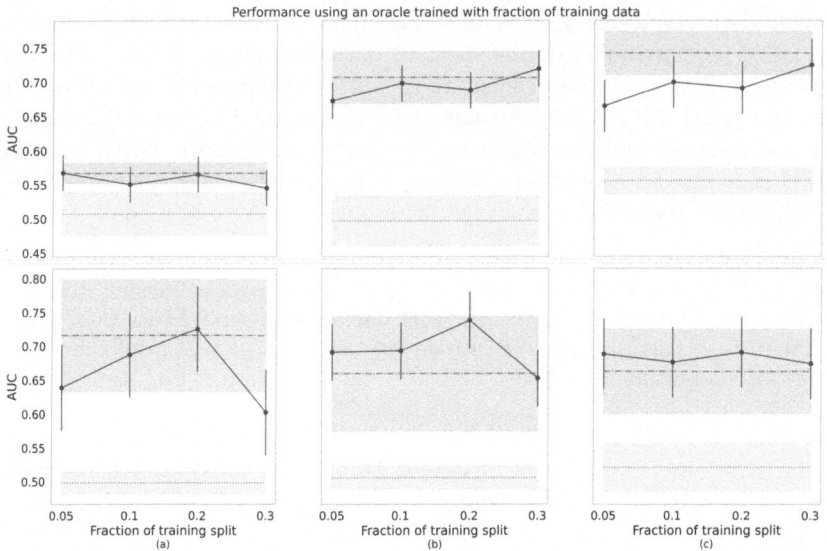

Fig. 4. AUC of a ResNet18 trained on: top: CheXpert; bottom: ISIC. Test sets contain: (a) noise; (b) constant square; (c) random square. Models are trained with an AdamW optimizer at a learning rate of $1{\times}e^{-3}$.

We remove all data used to train the oracle from the training split of the student model. This ensures that there is no data leakage between the two. Even with significantly reduced training data, our oracle-guided approach consistently outperforms the ERM baseline (Fig. 4). These results suggest our method can be effective even with a small, carefully curated, clean dataset. This reduces the burden on clinicians, as it is more feasible to ensure a smaller subset is free from shortcuts and biases. The method's efficacy with limited data enhances its practical applicability in real-world clinical settings where large, diverse, and unbiased datasets may be scarce.

5   Discussion and Conclusion

Our oracle-guided training approach mitigates shortcut learning by regulating a student model's learning across hidden layers, preventing reliance on shortcut features. Our approach not only addresses general shortcut learning but also has potential implications for fairness-related biases in medical AI systems. We demonstrate significant improvements to model robustness and generalizability across various datasets and architectures. Notably, success with reduced oracle training data (as low as 5% of the full dataset) suggests practical applicability in real-world scenarios where curating large, bias-free datasets is challenging.

We observe that ERM models trained with SGD consistently outperform those trained with AdamW when the training data contains shortcuts, indicating that SGD may be more robust against shortcuts. This aligns with recent research demonstrating SGD's superior generalization capabilities [29]. Additionally, our method's effectiveness varies across shortcut types and datasets. Notably, we see minimal improvement for the noise shortcut on the CheXpert dataset. We hypothesize that pneumothorax detection in chest X-rays relies heavily on textural information, and the noise shortcut might corrupt clinically meaningful features, impeding learning. This could explain the performance drop of our clean benchmark on the noise shortcut compared to localized shortcuts.

We acknowledge that screening for all possible shortcuts in the oracle's training data is a significant challenge, even in small datasets, due to the complex and subtle nature of shortcuts in medical image data. However, our promising results in Sect. 4.3 suggest that a non-optimal oracle, sufficiently well-calibrated to the downstream task, can still effectively guide the learning of a student on the full dataset. Furthermore, our approach offers significant potential in scenarios like continual learning, where a model needs to be updated over time without inadvertently introducing new shortcuts [15].

Future work should explore the application of our method to datasets with real-world shortcuts, as well as investigate its effectiveness in addressing fairness-related biases. Additionally, extension to the multi-class setting, using softmax class probabilities in place of sigmoid, would broaden its applicability. Finally, while we focused on ERM for training, examining the interplay between our method and alternative training techniques, such as Focal loss, or regularization and normalization methods, could potentially enhance model calibration and bias robustness [17, 25]. These explorations will help to further validate and expand our approach's utility across diverse scenarios in medical imaging AI.

Acknowledgments. This work was supported by the UKRI.

Disclosure of Interests. The authors have no competing interests to declare that are relevant to the content of this article.

References

1. Ahmadian, A., Lindsten, F.: Enhancing representation learning with deep classifiers in presence of shortcut. In: ICASSP 2023-2023 IEEE International Conference on Acoustics, Speech and Signal Processing (ICASSP), pp. 1–5. IEEE (2023)
2. Ahmed, K.B., Hall, L.O., Goldgof, D.B., Fogarty, R.: Achieving multisite generalization for CNN-based disease diagnosis models by mitigating shortcut learning. IEEE Access **10**, 78726–78738 (2022)
3. Ao, S., Rueger, S., Siddharthan, A.: Confidence-aware calibration and scoring functions for curriculum learning. In: Fifteenth International Conference on Machine Vision (ICMV 2022), vol. 12701, pp. 558–567. SPIE (2023)
4. Baldock, R.J.N., Maennel, H., Neyshabur, B.: Deep learning through the lens of example difficulty. In: Advances in Neural Information Processing Systems, vol. 34, pp. 10876–10889 (2021)
5. Boland, C., Goatman, K.A., Tsaftaris, S.A., Dahdouh, S.: There are no shortcuts to anywhere worth going: identifying shortcuts in deep learning models for medical image analysis. In: Medical Imaging with Deep Learning (2024)

6. Brown, A., Tomasev, N., Freyberg, J., Liu, Y., Karthikesalingam, A., Schrouff, J.: Detecting shortcut learning for fair medical AI using shortcut testing. Nat. Commun. **14**(1), 4314 (2023)
7. Cha, J., Lee, K., Park, S., Chun, S.: Domain generalization by mutual-information regularization with pre-trained models. In: Avidan, S., Brostow, G., Cissé, M., Farinella, G.M., Hassner, T. (eds.) European Conference on Computer Vision, pp. 440–457. Springer,Cham (2022). https://doi.org/10.1007/978-3-031-20050-2_26
8. Codella, N.C., et al.: Skin lesion analysis toward melanoma detection: a challenge at the 2017 international symposium on biomedical imaging (ISBI), hosted by the international skin imaging collaboration (ISIC). In: Proceedings - International Symposium on Biomedical Imaging, vol. 2018-April, pp. 168–172. IEEE Computer Society (2018)
9. Dagaev, N., Roads, B.D., Luo, X., Barry, D.N., Patil, K.R., Love, B.C.: A too-good-to-be-true prior to reduce shortcut reliance. Pattern Recogn. Lett. **166**, 164–171 (2023)
10. Du, Y., et al.: Less learn shortcut: analyzing and mitigating learning of spurious feature-label correlation. arXiv preprint arXiv:2205.12593 (2022)
11. Jiménez-Sánchez, A., Juodelyte, D., Chamberlain, B., Cheplygina, V.: Detecting shortcuts in medical images - a case study in chest x-rays. In: 2023 IEEE 20th International Symposium on Biomedical Imaging (ISBI), pp. 1–5 (2023)
12. Krizhevsky, A.: Learning multiple layers of features from tiny images (2009)
13. Lee, C.S., Lee, A.Y.: Clinical applications of continual learning machine learning. Lancet Digit. Health **2**(6), e279–e281 (2020)
14. Lee, D., Jung, S., Moon, T.: Issues for continual learning in the presence of dataset bias. In: Mundt, M., Cooper, K.W., Dhami, D.S., Ribeiro, A., Smith, J.S., Bellot, A., Hayes, T. (eds.) Proceedings of The First AAAI Bridge Program on Continual Causality. Proceedings of Machine Learning Research, vol. 208, pp. 92–99. PMLR (2023)
15. Lesort, T.: Continual feature selection: spurious features in continual learning. arXiv preprint arXiv:2203.01012 (2022)
16. Li, Y., Vasconcelos, N.: Repair: removing representation bias by dataset resampling. In: Proceedings of the IEEE/CVF Conference on Computer Vision and Pattern Recognition, pp. 9572–9581 (2019)
17. Lin, T.Y., Goyal, P., Girshick, R., He, K., Dollár, P.: Focal loss for dense object detection. In: Proceedings of the IEEE International Conference on Computer Vision, pp. 2980–2988 (2017)
18. Mei, X., et al.: RadImageNet: an open radiologic deep learning research dataset for effective transfer learning. Radiol. Artif. Intell. **4**(5), e210315 (2022)
19. Minderer, M., Bachem, O., Houlsby, N., Tschannen, M.: Automatic shortcut removal for self-supervised representation learning. In: International Conference on Machine Learning, pp. 6927–6937. PMLR (2020)
20. Murali, N., et al.: Shortcut learning through the lens of early training dynamics. arXiv:2302.09344 (2022)
21. Nauta, M., Walsh, R., Dubowski, A., Seifert, C.: Uncovering and correcting shortcut learning in machine learning models for skin cancer diagnosis. Diagnostics **12**, 40 (2021)
22. Sagawa, S., Koh, P.W., Hashimoto, T.B., Liang, P.: Distributionally robust neural networks for group shifts: on the importance of regularization for worst-case generalization. arXiv preprint arXiv:1911.08731 (2019)

23. Taha, A.A., Hennig, L., Knoth, P.: Confidence estimation of classification based on the distribution of the neural network output layer. arXiv preprint arXiv:2210.07745 (2022)
24. Utama, P.A., Moosavi, N.S., Gurevych, I.: Mind the trade-off: debiasing NLU models without degrading the in-distribution performance. arXiv preprint arXiv:2005.00315 (2020)
25. Wang, C.: Calibration in deep learning: a survey of the state-of-the-art. arXiv preprint arXiv:2308.01222 (2023)
26. Youssef, A., Abramoff, M., Char, D.: Is the algorithm good in a bad world, or has it learned to be bad? the ethical challenges of "locked" versus "continuously learning" and "autonomous" versus "assistive" AI tools in healthcare. Am. J. Bioeth. **23**(5), 43–45 (2023)
27. Zhang, R., Griner, D., Garrett, J.W., Qi, Z., Chen, G.H.: Training certified detectives to track down the intrinsic shortcuts in COVID-19 chest X-ray data sets. Sci. Rep. **13**(1), 12690 (2023)
28. Zhen, X., Meng, Z., Chakraborty, R., Singh, V.: On the versatile uses of partial distance correlation in deep learning. In: Avidan, S., Brostow, G., Cissé, M., Farinella, G.M., Hassner, T. (eds.) European Conference on Computer Vision, pp. 327–346. Springer, Cham (2022). https://doi.org/10.1007/978-3-031-19809-0_19
29. Zhou, P., et al.: Towards theoretically understanding why SGD generalizes better than Adam in deep learning. Adv. Neural. Inf. Process. Syst. **33**, 21285–21296 (2020)

Exploring Fairness in State-of-the-Art Pulmonary Nodule Detection Algorithms

John McCabe[1]([✉]), Daryl Cheng[1], Amyn Bhamani[2], Monica Mullin[2,5],
Tanya Patrick[2], Arjun Nair[4], Sam M. Janes[2], Carole H. Sudre[3],
and Joseph Jacob[1]

[1] Satsuma Lab, Centre for Medical Image Computing (CMIC),
University College London, London, UK
j.mccabe@ucl.ac.uk

[2] Lungs for Living Research Centre, UCL Respiratory, University College London,
London, UK

[3] Centre for Medical Image Computing (CMIC), University College London, London,
UK

[4] University College London Hospitals NHS Foundation Trust, London, UK

[5] Department of Respirology, University of British Columbia, Vancouver, Canada

Abstract. Lung cancer is the leading cause of cancer mortality worldwide. Asymptomatic in its early stages, it is disproportionately detected when the disease is advanced. Resource constraints have resulted in increasing reliance on computer-aided detection (CADe) systems to assist with scan evaluation. The datasets used to train these algorithms are often unbalanced in their representation of protected groups e.g. sex and ethnicity. This project investigates whether there are performance disparities in detecting clinically relevant nodules across under-represented groups in selected, state-of-the-art nodule detection algorithms trained on data from a screening program in the UK.

Our analysis revealed that overall, the algorithms demonstrate equitable performance across various demographic groups. However, their performance varies strongly across nodule characteristics (size and type) in line with their prevalence in the training set. To ensure continued equitable performance, algorithms should not only consider demographic but also nodule attributes representativeness in their training.

Keywords: Nodule Detection Algorithms · Fairness in AI · Lung Cancer Screening

1 Introduction

1.1 Background

Lung cancer is responsible for 20% of all cancer deaths, the largest number of cancer deaths worldwide [1]. The main reason for this is that lung cancer is often asymptomatic during the early stages. When it becomes symptomatic, the disease has invariably spread, preventing curative treatment and reducing survival.

E. Puyol-Antón et al. (Eds.): FAIMI 2024/EPIMI 2024, LNCS 15198, pp. 78–87, 2025.
https://doi.org/10.1007/978-3-031-72787-0_8

However, when lung cancer is detected in its early stages, effective treatments are increasingly emerging, improving survival rates. Five-year survival rates range between 10% and 36% for patients diagnosed at stages III and IV, but approach 53% to 92% in patients diagnosed at stages I and II [1].

A series of Randomized Control Trials (RCTs) [2,3] have demonstrated the effectiveness of using Low Dose Computed Tomography (LDCT) to identify lung cancer in high-risk populations, resulting in 20 to 24% reductions in lung cancer mortality. The main goal of an LDCT scan is to identify lung nodules. Lung nodules consist of a collection of cells, which together form a lesion large enough to be visualised at the spatial resolution of a LDCT scan. Some nodules may represent the early stages of a lung cancer. Accordingly, the most important task when assessing nodules on LDCT is to confidently detect nodules that are likely to harbour a cancer. These nodules require more detailed clinical investigation and are often termed actionable nodules. Detecting actionable nodules is a complex task as the majority of lung nodules are benign and do not need further clinical work-up. Similarly, many structures in the lung may mimic the appearance of nodules. The Fleischner Society has introduced terminology to standardize the descriptions and reporting of lung nodules [4], and also provides guidelines for follow-up and management of these nodules.

Alongside the previously described trials, numerous studies have been carried out to refine the lung screening process and assess the economic impact of implementing a screening program. As a result, the UK National Screening Committee has recommended targeted screening throughout all four UK nations, with plans for a complete roll out by 2029. This will result in a significant increase in demand for LDCT scans reporting. There is an existing shortage of qualified radiologists to carry out this work, with a projected shortfall of 39% by 2026 [5]. The pressures on radiological resources are likely to result in reporting delays and an overworked workforce, which will impact patient care, standards of practice and patient outcomes.

Artificial intelligence and machine learning (AI/ML) are being considered as a potential solution to these workforce challenges. Deep learning (DL), a subset of AI, has been at the forefront of the most significant developments in healthcare research in recent years. DL models have been successfully introduced into several healthcare settings such as a computer-aided detection (CADe) systems [6–8], drug discovery [9] and robotic surgery [10].

Multiple studies have demonstrated challenges related to the generalizability and bias in the development of DL models [11–13]. This includes the potential for bias in various medical imaging tasks, such as classification and segmentation, which are widely used used across different medical domains and diseases [14–16]. In the field of lung nodule detection, several commercial products have been licensed to act as second readers for lung cancer screening, assisting radiologists in interpreting lung nodules on LDCT scans. However, the specifics of the design and training of these commercial AI models are not publicly shared in order to protect intellectual property. These algorithms are likely to have been trained on publicly available nodule datasets. However, these datasets, including those from the aforementioned RCTs, have been shown to lack diversity (for example, the NELSON trial had a five to one male to female ratio).

The publicly available datasets utilized for developing nodule detection algorithms often consist of LDCT scans that are over a decade, and occasionally two decades old. These datasets may exhibit inconsistencies, particularly in the annotation of nodule types and sizes which encompass both actionable and non-actionable nodules.

There are three main aims of this project, first to understand whether there is any variation in performance across protected groups in current nodule detection algorithms. Secondly, if there are discrepancies, does this result from training with unbalanced datasets. Thirdly, can we determine what the main drivers of performance are within nodule detection algorithms.

2   Methods and Materials

2.1   Study Design

This study employed a comparative analysis approach to evaluate performance variation in nodule detection algorithms across sex and ethnic group. Two state-of-the-art nodule detection algorithms were chosen for assessment based on availability, diversity in architecture and performance. The evaluation was conducted using a dataset drawn from the SUMMIT study, one of Europe's largest lung cancer screening programs.

2.2   Dataset Description

The SUMMIT cohort is a London based lung screening study involving high-risk participants invited from primary care. Participants were risk assessed, and qualifying participants were consented to undergo an LDCT scan and provide indication of sex and ethnic group. For this analysis, a subset of 5,290 baseline LDCT scans was utilized. These scans, which were the ones available at the start of the project, were randomly selected and closely mirror the demographic composition of the overall SUMMIT cohort. The SUMMIT cohort itself exhibits an imbalance, with men slightly outnumbering women and the 'White' ethnic group significantly outnumbering all other ethnicity's in the sample. Additionally, a substantial sex imbalance was observed within 'Asian or Asian British' and 'Black' ethnic groups, with males greatly outnumbering females.

Each LDCT was reported by a specialist pulmonary radiologist, supported by the Mevis'™ Veolity CADe software [6]. In addition to the location of the nodule, other characteristics including maximum diameter and type (solid, part-solid, non-solid, consolidation) were recorded.

2.3   Nodule Detection Algorithms

Two object detection open-source frameworks with different backbones and pre-processing and having demonstrated state-of-the-art performance on the

LUNA16 [17] nodule detection challenge were considered to assess the impact of model architecture on the study findings.

Model 1 is the winning entry in the Kaggle Data Science Bowl 2017 [18]. This model consists of a one-stage object detector utilising a modified UNet [19] architecture as backbone with approximately 5 million trainable parameters. The pre-processing include a lung segmentation step, resampling and intensity clipping (−1200 to 300 Hounsfield Units (HU)). Following the original training strategy used to develop this algorithm, nodules with a diameter greater than 30 mm were oversampled during training

Model 2 is the MONAI detection algorithm which utilizes a RetinaNet [20] with approximately 21 million trainable parameters. This uses a feature pyramid network [21] to enable detection at different scales and a Focal Loss [20] to deal with class imbalance. Pre-processing includes a resampling and a clipping step (−1024 to 300 HU).

Both models use hard-negative mining [22] and a patch-based training method (128^3 and $198 \times 198 \times 80$ for Model 1 and 2 respectively).

2.4 Evaluation Metrics

Given their implication for patient management, evaluation focused on detection performance for actionable nodules, i.e. nodules for which specific follow-up is required. Performance was assessed using the Free-Response Receiver Operating Characteristic (FROC) curves, which measures sensitivity over 7 fixed false positives per scan operating points ($\frac{1}{8}$, $\frac{1}{4}$, $\frac{1}{2}$, 1, 2, 4, 8) and a Competition Performance Metric (CPM), which is the average sensitivity over these operating points. Bootstrap (1000 samples) was used to obtain confidence intervals.

2.5 Experiments

Experiments were conducted to evaluate the impact of training set imbalance as observed in the make-up of the screening cohort on selected nodule detection algorithms with a focus on differences across sex and ethnic group. For comparative purposes and across all experiments, all test groups were balanced and the training set was built to mimic the distribution of the screening sample. Due to their low number and their heterogeneity, "Other" and "Mixed" ethnic groups were only considered in the training set. Experiments can be described as follows

- **Experiment 1** - Maximisation of training set size given minimum test set ethnic group size.
- **Experiment 2** Isolation of the impact of ethnic group on performance without confounding for sex by training and testing on male samples only.
- **Experiment 3** Isolation of the impact of sex imbalance on performance without confounding for ethnicity by training and testing on white only.
- **Experiment 4** Comparison of performance across nodule types and sizes to understand drivers of performance using settings of Experiment 1.

Table 1 shows the composition of experiments 1-3. The proportions of sex and ethnic group for the whole SUMMIT sample are shown in the end column.

Table 1. Profile of protected groups for Training, Validation and Test datasets used for each experiment

Protected Group	Category	Experiment 1			Experiment 2 - Male only			Experiment 3 - White only			SUMMIT
		Training	Validation	Test	Training	Validation	Test	Training	Validation	Test	Total
Sex	Female	1961 (38.6%)	125 (46.8%)	250 (42.1%)	0 (0%)	0 (0%)	0 (0%)	1494 (44.4%)	98 (43.8%)	399 (50.0%)	5508 (42.5%)
	Male	3118 (61.4%)	142 (53.2%)	344 (57.9%)	1573 (100.0%)	105 (100.0%)	420 (100.0%)	1870 (55.6%)	126 (56.2%)	399 (50.0%)	7450 (57.5%)
Ethnic Group	Asian or Asian British	443 (8.7%)	25 (9.4%)	198 (33.3%)	98 (6.2%)	11 (10.5%)	140 (33.3%)	0 (0%)	0 (0%)	0 (0%)	845 (6.5%)
	Black	244 (4.8%)	12 (4.5%)	198 (33.3%)	71 (4.5%)	4 (3.8%)	140 (33.3%)	0 (0%)	0 (0%)	0 (0%)	577 (4.5%)
	Mixed	119 (2.3%)	9 (3.4%)	0 (0%)	34 (2.2%)	2 (1.9%)	0 (0%)	0 (0%)	0 (0%)	0 (0%)	283 (2.2%)
	Other ethnic groups	199 (3.9%)	7 (2.6%)	0 (0%)	53 (3.4%)	4 (3.8%)	0 (0%)	0 (0%)	0 (0%)	0 (0%)	451 (3.5%)
	White	4074 (80.2%)	214 (80.1%)	198 (33.3%)	1317 (83.7%)	84 (80.0%)	140 (33.3%)	3364 (100.0%)	224 (100.0%)	798 (100.0%)	10802 (83.4%)
Total		5079	267	594	1573	105	420	3364	224	798	12958

3   Results

Results for experiments 1-3 are presented as bar plots of the mean sensitivity and 95% confidence interval at the seven fixed operating points across protected groups for each model.

3.1   Experiment 1: Test Dataset with Balanced Ethnic Groups

The results for Experiment 1 are shown in Fig. 1. The first row (Fig. 1a and 1b) presents the outcomes from Model 1. The plots suggest similar performance between male and female participants, although female participants show a higher CPM of 0.46 (95% CI 0.38-0.55) compared to 0.38 (95% CI 0.30-0.46) in male participants.

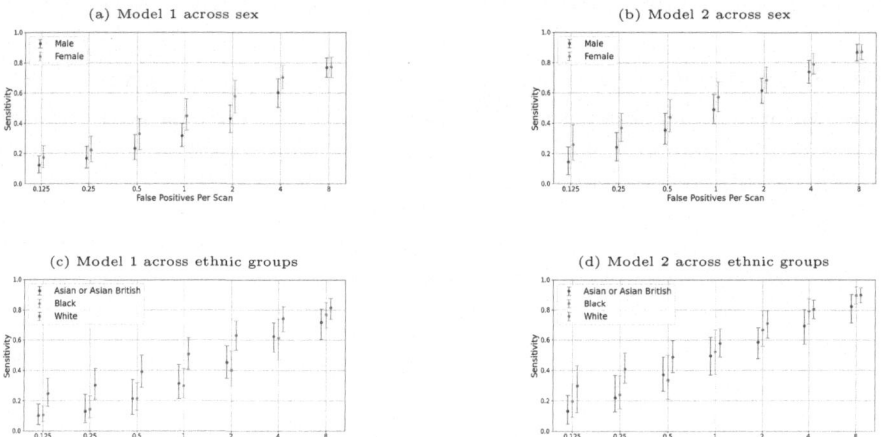

Fig. 1. Sensitivity bar plots (mean 95% CI) across sex (top row) and ethnic groups (bottom row) for the two models trained on SUMMIT dataset.

When comparing ethnic groups in Model 1, 'White' participants exhibit a higher CPM of 0.52 (95% CI 0.43-0.61) compared to 'Asian or Asian British' and 'Black' participants, who show CPMs of 0.36 (95% CI 0.27-0.47) and 0.36 (95% CI 0.28-0.46) respectively.

The results from Model 2, depicted in the second row (Fig. 1c and 1d), generally indicate better performance across all categories compared to Model 1. A similar pattern is observed where females in Model 2 having a higher CPM of 0.57 (95% CI 0.49-0.66) compared to males, who show a CPM of 0.49 (95% CI 0.41-0.58). Additionally, 'White' participants in Model 2 achieve a CPM of 0.60 (95% CI 0.50-0.69), which is better than 'Asian or Asian British' participants with a CPM of 0.47 (95% CI 0.36-0.59) and 'Black' participants with a CPM of 0.52 (95% CI 0.42-0.64).

3.2   Experiment 2: Male Only

Figure 2 shows the results for each ethnic group when trained on a male-only sample. For both Model 1 and Model 2, performance across ethnic groups exhibits minimal variation. In Model 1, the CPM varies across different ethnic groups: White participants exhibit a Mean Sensitivity of 0.43 (95% CI 0.33-0.54), 'Asian or Asian British' participants show a Mean Sensitivity of 0.48 (95% CI 0.35-0.62), and 'Black' participants demonstrate a Mean Sensitivity of 0.41 (95% CI 0.26-0.578). For Model 2, the CPMs are as follows: White participants have a Mean Sensitivity of 0.53 (95% CI 0.42-0.653), 'Asian or Asian British' participants show a Mean Sensitivity of 0.58 (95% CI 0.45-0.719), and 'Black' participants exhibit a Mean Sensitivity of 0.49 (95% CI 0.37-0.627). It should be noted that the mean sensitivity shifts between these groups across various operating points for both models and interestingly, the 'Asian or Asian British' participants, who are under-represented in the sample perform marginally better for both models.

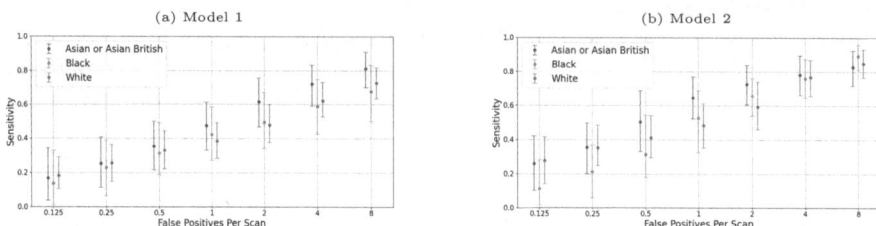

Fig. 2. Sensitivity bar plots (mean 95% CI) across ethnic groups at fixed operating points for models trained on a male-only sub-sample.

3.3   Experiment 3: White Only

The results for Experiment 3, as shown in Fig. 3, indicate that contrary to Experiment 1, where females had a higher CPM, when trained exclusively on white participants there are only very small differences in CPM. Model 1 shows that male participants have a CPM of 0.44 (95% CI 0.38-0.509), while female participants exhibit a CPM of 0.44 (95% CI 0.37-0.51). For Model 2 in Experiment 3, male participants demonstrate a CPM of 0.49 (95% CI 0.42-0.557), whereas female participants show a Mean Sensitivity of 0.51 (95% CI 0.45-0.574).

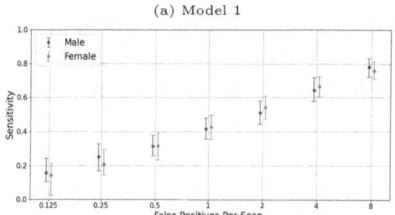 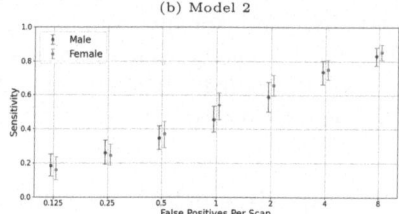

Fig. 3. Sensitivity bar plots (mean 95% CI) across sex at fixed operating points for models trained on White participants only

3.4   Experiment 4

The performance across each nodule size and type at the different operating points for the two models trained on data from Experiment 1 is presented in Fig. 4. Regarding the diameter plots, the contrasting algorithm designs are apparent: Model 1, employing over-sampling for nodules sized 30–40 mm and 40+ mm, detects larger nodules at lower operating points. In contrast, Model 2 detects a greater proportion of smaller nodules at lower operating points, possibly benefiting from the scale robustness provided by the FPN. Regarding the nodule types, Model 1 shows earlier detection of part-solid nodule types compared to Model 2. Both models demonstrate similarly higher performance in detecting the most frequently found nodules in the training dataset (solid nodules) and lower performance in detecting less common nodules (non-solid and consolidation types)

4   Discussion

In this study, the variation in performance across sex and ethnic groups was evaluated in two state-of-the-art pulmonary nodule detection algorithms trained on an unbalanced dataset drawn from a large screening program. There was no indication of adverse impact of unbalanced demographic representation on nodule detection performance when addressing confounding factors.

When the prevalence of different nodule types were evaluated in the training dataset marked differences in actionable nodule detection were uncovered. The most common nodule subtypes were generally better detected than their rarer counterparts. The one exception to this was for part-solid nodules. Though a rare nodule subtype, algorithm performance for detecting part-solid nodules was

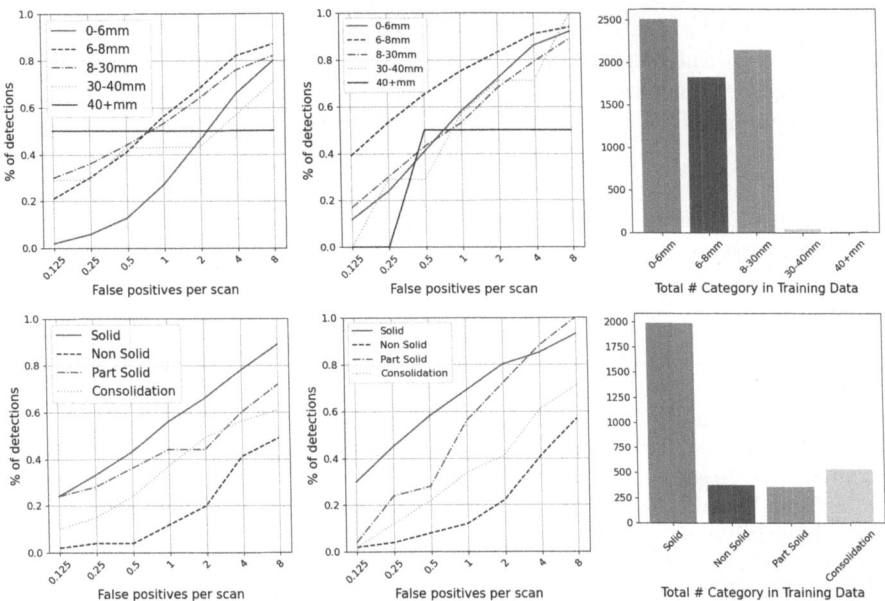

Fig. 4. Sensitivity for each nodule category by size (top row) and nodule type (bottom row) for Model 1 (left) and Model 2 (right) at the fixed operating points. Training proportions are indicated in the bar plot on the right.

notably high, potentially due to shared features in a part-solid nodule with the much more commonly found solid nodule. As nodule type prevalence varies across different lung cancer screening populations, there may be a class imbalance in training datasets fed to an AI algorithm. These algorithms may then perform poorly when detecting certain important nodule subtypes. It is therefore essential to design robust training strategies that take account of and cohort enrich training datasets for the various important and potentially under-represented nodule classes. Such strategies should prevent potential cancer being missed by an AI algorithm. Model design appeared to be an influential factor determining algorithm performance with respect to the size of the nodules. Such findings also underline the need to carefully reconsider the clinical relevance of the evaluation metrics used for comparison. In practice, detection at low operating points for nodules with a size above 8mm may be a more appropriate focus of attention.

This study is limited in its generalizability by the implications of the screening setting limited to a single scanner type with a consistent protocol that is not representative of clinical setting. Future research should expand the understanding of performance difference across nodule characteristics (e.g. location), extend to not screened populations (e.g. non-smokers) and elaborate on metrics more suitable for clinical application (e.g. focus on low operating points).

In conclusion, our analysis demonstrates reassuring results regarding the overall fairness of existing nodule detection algorithms. However, it also underscores the need for continued efforts to ensure sustained performance not only across diverse demographic populations but also nodule subtypes presentations. It challenges the one-size-fits-all approaches in nodule detection algorithms and promotes the need for the development of solutions tailored to nodule characteristics.

Disclosure of Interests. The authors declare that they have no known competing financial interests or personal relationships that could have appeared to influence the work reported in this paper.

References

1. Cancer Statistics for the UK [WWW Document]. Cancer Research UK (2015). https://www.cancerresearchuk.org/health-professional/cancer-statistics-for-the-uk. Accessed 23 Mar 2023
2. National Lung Screening Trial Research Team, et al.: The national lung screening trial: overview and study design. Radiology **258**, 243 (2011). https://doi.org/10.1148/radiol.10091808
3. de Koning, H.J., et al.: Reduced lung-cancer mortality with volume CT screening in a randomized trial. New England J. Med. **382**, 503–513 (2020). https://doi.org/10.1056/NEJMoa1911793
4. Bankier, A.A., et al.: Fleischner society: glossary of terms for thoracic imaging. Radiology **310**, e232558 (2024). https://doi.org/10.1148/radiol.232558
5. RCR Clinical radiology census report 2021 | The Royal College of Radiologists [WWW Document]. https://www.rcr.ac.uk/clinical-radiology/rcr-clinical-radiology-census-report-2021. Accessed 23 Mar 2023
6. Veolity - a brand of MeVis Medical Solutions AG: product information [WWW Document]. https://www.veolity.com/about-veolity/product-information. Accessed 23 Mar 2024
7. Veye Lung Nodules [WWW Document], Aidence. https://www.aidence.com/veye-lung-nodules/. Accessed 23 Mar 2024
8. AI-Rad Companion [WWW Document]. https://www.siemenshealthineers.com/en-uk/digital-health-solutions/ai-rad-companion. Accessed 23 Mar 2024
9. Yang, X., Wang, Y., Byrne, R., Schneider, G., Yang, S.: Concepts of artificial intelligence for computer-assisted drug discovery. Chem. Rev. **119**, 10520–10594 (2019). https://doi.org/10.1021/acs.chemrev.8b00728
10. Beyaz, S.: A brief history of artificial intelligence and robotic surgery in orthopedics & traumatology and future expectations. Jt. Dis. Relat. Surg. **31**, 653–655 (2020). https://doi.org/10.5606/ehc.2020.75300
11. Buolamwini, J., Gebru, T.: Gender shades: intersectional accuracy disparities in commercial gender classification. In: Proceedings of the 1st Conference on Fairness, Accountability and Transparency, pp. 77–91. PMLR (2018)
12. Wang, M., Deng, W.: Mitigate bias in face recognition using skewness-aware reinforcement learning. arXiv:1911.10692 (2019)
13. Brandao, M.: Age and gender bias in pedestrian detection algorithms. arXiv:1906.10490 (2019)
14. Puyol-Anton, E., et al.: Fairness in cardiac MR image analysis: an investigation of bias due to data imbalance in deep learning based segmentation. arXiv:2106.12387 (2021)

15. Weng, N., Bigdeli, S., Petersen, E., Feragen, A.: Are sex-based physiological differences the cause of gender bias for chest X-ray diagnosis? arXiv:2308.05129 (2023)
16. Burlina, P., Joshi, N., Paul, W., Pacheco, K.D., Bressler, N.M.: Addressing artificial intelligence bias in retinal diagnostics. Transl. Vis. Sci. Technol. **10**, 13 (2021). https://doi.org/10.1167/tvst.10.2.13
17. LUNA16 - Grand Challenge. https://luna16.grand-challenge.org/Data/. Accessed 18 Apr 2024
18. Liao, F., Liang, M., Li, Z., Hu, X., Song, S.: Evaluate the malignancy of pulmonary nodules using the 3D deep leaky noisy-or network. IEEE Trans. Neural Netw. Learn. Syst. **30**, 3484–3495 (2019). https://doi.org/10.1109/TNNLS.2019.2892409
19. Ronneberger, O., Fischer, P., Brox, T.: U-Net: convolutional networks for biomedical image segmentation. arXiv:1505.04597 (2015)
20. Lin, T.-Y., Goyal, P., Girshick, R., He, K., Dollár, P.: Focal loss for dense object detection. arXiv:1708.02002 (2018)
21. Lin, T.-Y., Dollár, P., Girshick, R., He, K., Hariharan, B., Belongie, S.: Feature pyramid networks for object detection. arXiv:1612.03144 (2017)
22. Bucher, M., Herbin, S., Jurie, F.: Hard negative mining for metric learning based zero-shot classification. arXiv:1608.07441 (2016)

Quantifying the Impact of Population Shift Across Age and Sex for Abdominal Organ Segmentation

Kate Čevora[1]([✉]), Ben Glocker[1], and Wenjia Bai[1,2,3]

[1] Department of Computing, Imperial College London, London, UK
{kc2322,b.glocker,w.bai}@imperial.ac.uk
[2] Department of Brain Sciences, Imperial College London, London, UK
[3] Data Science Institute, Imperial College London, London, UK

Abstract. Deep learning-based medical image segmentation has seen tremendous progress over the last decade, but there is still relatively little transfer into clinical practice. One of the main barriers is the challenge of domain generalisation, which requires segmentation models to maintain high performance across a wide distribution of image data. This challenge is amplified by the many factors that contribute to the diverse appearance of medical images, such as acquisition conditions and patient characteristics. The impact of shifting patient characteristics such as age and sex on segmentation performance remains relatively under-studied, especially for abdominal organs, despite that this is crucial for ensuring the fairness of the segmentation model. We perform the first study to determine the impact of population shift with respect to age and sex on abdominal CT image segmentation, by leveraging two large public datasets, and introduce a novel metric to quantify the impact. We find that population shift is a challenge similar in magnitude to cross-dataset shift for abdominal organ segmentation, and that the effect is asymmetric and dataset-dependent. We conclude that dataset diversity in terms of known patient characteristics is not necessarily equivalent to dataset diversity in terms of image features. This implies that simple population matching to ensure good generalisation and fairness may be insufficient, and we recommend that fairness research should be directed towards better understanding and quantifying medical image dataset diversity in terms of performance-relevant characteristics such as organ morphology.

Keywords: Abdominal CT Segmentation · Generalisation · Fairness

1 Introduction

Automated medical image segmentation models have seen tremendous progress in terms of segmentation speed and accuracy, in some cases surpassing the performance of human experts [8,11,22]. However, there is a large gap at present

---

Supplementary Information The online version contains supplementary material available at https://doi.org/10.1007/978-3-031-72787-0_9.

© The Author(s), under exclusive license to Springer Nature Switzerland AG 2025
E. Puyol-Antón et al. (Eds.): FAIMI 2024/EPIMI 2024, LNCS 15198, pp. 88–97, 2025.
https://doi.org/10.1007/978-3-031-72787-0_9

between the plethora of automated segmentation models which are developed in research environments, and those which are integrated into clinical practice. A commonly cited reason for this gap is the often poor generalisation performance of segmentation models to test data which is outside of the distribution of the training data, known as domain shift [28].

When we look at domain shift in medical image segmentation via the lens of causality [2], three common types of shift exist, namely population shift, acquisition shift and annotation shift, illustrated by the casual graph in Fig. 1. Population shift is caused by changes in the distribution of patient characteristics such as age, sex, ethnicity and disease prevalence [27]. It is particularly important because it has the potential to result in biased model predictions for different patient populations. While acquisition and annotation shift have received significant attention leading to a range of advanced augmentation approaches, domain adaptation methods [3] and standard operating procedures for annotation [23] to mitigate their effects, population shift receives relatively less research attention, in particular for abdominal organ segmentation.

To better understand the influence of population shift on abdominal organ segmentation, we collate a large-scale abdominal CT dataset of 1,582 subjects from public sources along with their population characteristics. We perform the first study to evaluate the impact of population shift with respect to age and sex on segmentation performance for major abdominal organs, and introduce a novel metric, the performance gap, to quantify the maximal impact of population shift for each subgroup of interest. We also compare the impact of population shift on segmentation performance to that caused by cross-dataset shift. Furthermore, we propose a novel hypothesis that the segmentation performance is more directly determined by the training set diversity in terms of image features, rather than population characteristics. We believe that our findings, the evaluation framework and our recommendations for the direction of future research will provide useful insights for the community to elucidate the complex causes and magnitude of population shift in medical image segmentation problems.

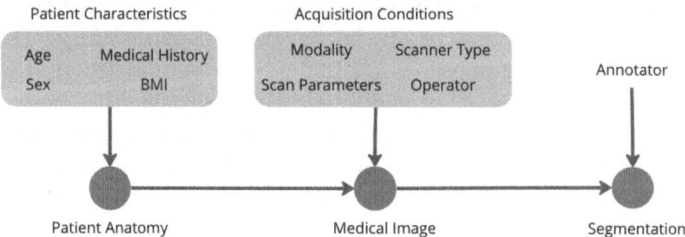

Fig. 1. Causal diagram illustrating major factors that can influence medical image appearance and associated segmentation. The factors can be split into three broad groups: patient characteristics which directly influence patient anatomy, acquisition conditions which influence image appearance, and annotation protocol which influences manual segmentation style.

2   Background and Related Works

Domain Shift is a significant challenge for medical image segmentation, occurring when there is a significant shift in the statistical distribution of the appearance of medical imaging data across different sources. Figure 1 shows a simplified causal perspective on the factors contributing to image appearance and corresponding segmentation, which can be broadly split into three groups: patient characteristics, acquisition conditions and annotation style. Changes in these factors manifest in medical images in the form of differing anatomical shapes, contrasts, intensity distribution, resolution, or noise patterns. As a result, segmentation models trained on one dataset may not generalise well across data from different sources [9,12,20].

Population Shift is a specific type of domain shift which is caused by changes in the relative proportion of subgroups in a dataset [15]. In the context of medical image datasets, subgroups are generally defined by patient characteristics such as age, sex, ethnicity or medical history. Several recent works demonstrate bias in image classification models arising from population shift with respect to sex and ethnicity [6,17,25]. This is particularly concerning because under-performance on certain populations at test-time can potentially lead to worse health outcomes for these groups.

There are relatively fewer works examining the impact of population shift on segmentation performance. Ioannou et al. [10] find significant race and sex bias with respect to accuracy for segmentation models trained on unbalanced brain imaging datasets. Lee et al. [18] found that segmentation models trained on cardiac MR images performed worse on racial groups which were underrepresented in the training data.

Remaining Challenges: Despite evidence that population shift can have a significant impact on the performance of medical image segmentation models [10,18], it is relatively under-studied compared to the impact of acquisition shift. For example, we are unaware of any other works that investigate the impact of population shift with respect to age and sex on segmentation of abdominal organs. Further, for organs and modalities where this impact has been quantified [10,18], the underlying causal mechanism of this bias has not been investigated. Gaining an understanding of the mechanisms of how population shift leads to change of performance is crucial for designing methods, such as data augmentation strategies, to mitigate its potential impact.

3   Method

3.1   Data

Although numerous efforts have been devoted to curating large-scale abdominal CT datasets [19], most of them do not release patient characteristics. After communicating with the owners of 13 public abdominal CT datasets, we were able to obtain patient-level demographic information for three. Two of them, TotalSegmentator (TS) [26] and AMOS [13], were sufficiently large to allow sex- and age-based resampling of training datasets to investigate the impact of population shift, which we will use for this work. Further details about the datasets are included in the Supple. Table 1, and will be released with the paper.

3.2 Experimental Design

We investigate the effects of population shift with respect to sex and age on segmentation performance for four abdominal organs: the left and right kidneys, pancreas and liver. Changes in shape and composition of these organs with respect to sex and age are known to occur [4,5,14,16,21,24,26,29], making them interesting candidates for investigation. Additionally, we perform a cross-dataset shift experiment to understand the magnitude of population shift in comparison to cross-dataset shift, the latter being significantly better-studied in the domain generalisation literature [7,28].

Measuring the Impact of Population Shift: We construct two subgroups, g_1 and g_2, for each patient characteristic (sex or age) by sampling without replacement from the full dataset (TotalSegmentator or AMOS). For sex, one subgroup contains only male subjects and the other contains only female. For age, one subgroup contains only subjects under 50 years old and the other contains only subjects over 70 years old. Each subgroup is further split into training and test sets. We train a segmentation model using the training set from a single subgroup, and then evaluate the trained model on the test sets from both subgroups.

To quantify the impact of population shift, we propose a new metric, the *performance gap* ΔP, which measures the change of segmentation performance, e.g. Dice score or 95-percentile Hausdorff distance (HD95), caused by the maximal shift of training set characteristics. The performance gap is normalised by the average segmentation performance and formulated as,

$$\Delta P_{g_1}(g_1, g_2) = \frac{P(g_1, S(g_2)) - P(g_1, S(g_1))}{0.5 \times [P(g_1, S(g_1)) + P(g_1, S(g_2))]} \times 100\% \qquad (1)$$

where $P(g_1, S(g_1))$ denotes the performance of a segmentation model S trained on subgroup g_1 and tested on subgroup g_1, $P(g_1, S(g_2))$ denotes the performance of a model trained on subgroup g_2 and tested on subgroup g_1, and ΔP_{g_1} denotes their performance gap when deployed on subgroup g_1. Similarly, we can define the performance gap ΔP_{g_2} when deployed on subgroup g_2.

The significance of a performance gap is calculated as a t-test carried out between $P(g_1, S(g_1))$ and $P(g_1, S(g_2))$.

Measuring the Impact of Cross-Dataset Shift: To understand the magnitude of population shift compared to other major sources of domain shift, we investigate the impact of cross-dataset shift. We construct two subgroups sampled from the TotalSegmentator [26] and AMOS [13] datasets respectively. We control for sex and age so that the two subgroups have similar population distributions, meaning that the remaining sources of shift between the two subgroups are mainly scanner, site, study type and disease type. We train segmentation models and assess the performance gap under cross-dataset shift using the same definition Eq. (1), where g_1 is formed of subjects from AMOS, and g_2 is formed of subjects from TotalSegmentator.

Measuring Training Data Diversity: To measure the diversity of the training data, we define a proxy measure of diversity, using the standard deviation of the organ volumes calculated across the training subjects in each subgroup.

Implementation Detail: We use a state-of-the-art 3D nnU-Net [11] as the segmentation model, although other architectures can also be used. nnU-Net appears regularly in the leaderboards of recent medical image segmentation challenges [1,13], and it has an established image pre-processing and data augmentation pipeline. For fair comparison, we ensure that the training set size is the same for both subgroups of a given dataset. Training set sizes for each experiment can be found in the Supple. Table 4. The validation set for parameter tuning is automatically selected by nnU-Net from the training samples. We employ 5-fold cross validation with a hold-out test set for each experiment and report the average results across the folds.

4 Results

Table 1 reports the observed performance gaps per dataset, organ and subgroup, along with the results for cross-dataset shift. A green fill indicates significant better performance when the test set matches the training set (positive value for Dice, negative value for HD95) and a red fill indicates significant worse performance when the test set matches the training set (negative value for Dice, positive value for HD95). Figure 2 shows the test set performance in terms of average Dice plotted against the organ volume diversity of the training dataset, split by subgroup. Raw average Dice scores for each experiment can be found in Supple. Table 3. Below we summarise the main findings:

The Impact of Population Shift is Significant for Kidney Segmentation. We see significant performance gaps in terms of both Dice and HD95 for the kidneys under population shift with respect to age and sex. This gap is particularly large for the male kidneys, where we see a performance drop of around 6% for Dice, and 95% for HD95. The magnitude of the significant performance gaps across organs observed for population shift (1–6% Dice, 95–125% HD95) is similar to that observed for cross-dataset shift (5–11% Dice, 100–135% HD95).

Table 1. Performance gaps ΔP in terms of Dice score and 95 percentile Hausdorff distance (HD95) due to population shift and cross-dataset shift. Coloured cells indicate that the performance gap is statistically significant ($p < 0.05$) via a t-test (N = group size, see Supple. Table 4), with red indicating a negative performance gap and green indicating a positive performance gap. Note that for Dice, a negative value indicates worse performance when the training set does not match the test set and for HD95, this is indicated by a positive value. TS: TotalSegmentator; U50: under 50; O70: over 70.

Dataset	Organ	ΔP, **Dice** (%)		ΔP, **HD95** (%)	
		$g_1 =$ **Female**	$g_2 =$ **Male**	$g_1 =$ **Female**	$g_2 =$ **Male**
TS	R. kidney	3.57	−5.94	20.5	37.1
	L. kidney	2.45	−6.17	−10.6	95.3
	Liver	1.61	−0.67	21.7	23.1
	Pancreas	4.15	−2.79	−10.4	11.2
AMOS	R. kidney	0.27	−0.11	−14.3	−2.13
	L. kidney	1.25	−0.42	−119.3	89.8
	Liver	−2.63	−0.23	−8.7	−30.7
	Pancreas	1.40	−1.64	−22.3	−7.5
Dataset	**Organ**	$g_1 =$ **U50**	$g_2 =$ **O70**	$g_1 =$ **U50**	$g_2 =$ **O70**
TS	R. kidney	−0.38	0.19	62.4	89.0
	L. kidney	1.65	−1.67	−124.7	108.8
	Liver	−0.87	0.18	12.4	4.8
	Pancreas	1.11	3.10	−10.2	29.1
AMOS	R. kidney	0.48	−0.23	−42.4	−52.9
	L. kidney	1.04	−0.25	−132.7	−5.2
	Liver	−0.72	−0.67	41.3	−1.1
	Pancreas	0.44	−1.99	−0.6	4.3
Dataset	**Organ**	$g_1 =$ **AMOS**	$g_2 =$ **TS**	$g_1 =$ **AMOS**	$g_2 =$ **TS**
TS/AMOS	R. kidney	0.46	−1.12	16.7	134.1
	L. kidney	−3.57	−3.90	135.2	112.1
	Liver	0.41	−4.66	66.3	151.4
	Pancreas	0.24	−10.7	29.7	101.3

Proportionate Representation of Subgroups Defined by Age and Sex is Not Sufficient to Ensure the Best Performance for These Groups at Test-Time.

Our results show that in some cases, a complete lack of representation in the training data can surprisingly result in better test-time performance compared to when the training and test data match in terms of population characteristics. For the female subgroup sampled from the AMOS dataset, test time performance on the left kidney is significantly better in terms of Dice (1.25%)

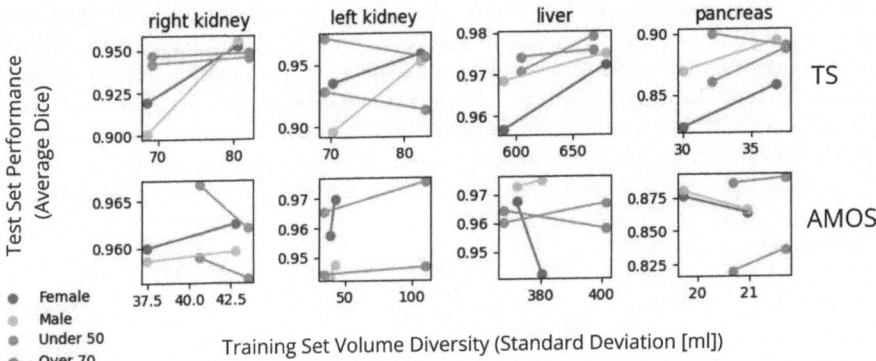

Fig. 2. Plots of segmentation performance in terms of Dice score on the test set against the proxy measure of training set diversity, the standard deviation of organ volumes. The test set data has been split by colour-coded subgroups. The top row reports results on the TotalSegmentator dataset (TS) and the bottom row reports results on AMOS.

and HD95 (−119%) when the training dataset is fully male, compared to when it is fully female. This is also true for the left kidney in the under 50 subgroup for both AMOS and TotalSegmentator (TS).

Proxy Measure of Training Data Diversity May Positively Correlate with Segmentation Performance. Figure 2 shows that increased diversity in the training dataset in terms of organ volume standard deviation correlates with increased test-set segmentation performance, in particular on the TotalSegmentator dataset, and possibly for the left and right kidneys on the AMOS dataset. Statistics per subgroup for training the set can be found in Supp. Tab. 2.

The Performance Gap is Asymmetric Between Subgroups. In cases where we see a significant performance gap for one subgroup (e.g. TS male kidneys), the complementary subgroup (TS female kidneys) does not necessarily show a similar performance gap. The TS male kidneys have a larger standard deviation of volumes compared with the female subgroup (80mL compared to 69mL), indicating greater diversity, which may explain this asymmetric performance gap. This asymmetry can also be observed in the cross-dataset shift experiments, where training a model with AMOS images causes a significant drop in performance for TS test images, but the same is not true for the AMOS test images.

5   Discussion and Future Directions

This is the first study quantifying the potential impact of population shift with respect to age and sex on the performance of abdominal CT image segmentation, using a state-of-the-art models with a standard set of image augmentations. Our results demonstrate that the impact of population shift with respect to age and sex is significant and can be comparable in magnitude to that caused by cross-dataset shift. This implies that the standard image augmentations employed by many image pre-processing pipelines such as rotation, scaling and random deformation, are insufficient to mitigate these effects. In order to simulate truly diverse abdominal CT datasets, we likely need more advanced image augmentation methods which can simulate real morphological differences between subgroups.

A common and well-supported hypothesis is that under-representation of a subgroup in training data can lead to decreased performance at test-time [10,17,18]. However, we have observed that for female kidneys, test-time segmentation performance could be improved by using a male training dataset. These findings are important because not only do they demonstrate potential for bias against certain groups under population shift, they imply that population-matching between training and test data is not sufficient to ensure fairness.

We hypothesise that this outcome may be the result of an imperfect correlation between diversity in terms of patient labels (e.g. sex, age, ethnicity) and diversity in terms of raw image features such as organ morphology, volume and texture. For example, we have observed in this case the male training images showed greater diversity in terms of organ volume than the female dataset, which may explain why the male-trained segmentation model showed better generalisation ability, even outperforming a female-trained model on female images.

We conclude that the impact of population shift with respect to age and sex on performance is significant for abdominal CT segmentation. Proportionate representation of subgroups defined by age and sex is not sufficient to ensure equitable performance at test-time. An initial look at the correlation between training dataset diversity in terms of organ volumes and segmentation performance suggests that measurements of diversity derived from raw image features are likely an important indicator of generalisation performance across subgroups.

In terms of future directions, our findings call for the development of methods to understand and measure medical image dataset diversity directly from raw image-level features such as shape, texture and volume. Such a metric will allow us to build training datasets and design image augmentation methods for medical image segmentations that result in better generalisation across a range of subgroups, without requiring per-patient demographic information. It will also enable predictions of whether a particular dataset is likely to result in a trained segmentation model that shows good test-time generalisation.

Limitations: We have attempted to control the effect of other potentially confounding variables (such as acquisition site, scanner type and study type) on our results by matching distributions of these variables as closely as possible between paired subgroups. However, successfully studying the effect of just a single variable in isolation on segmentation performance is a near-impossible task. Whilst it is theoretically possible to control for some known potentially confounding variables when designing experiments, many more are unknown or unreported. This aligns with our recommendation that fairness research in medical image analysis should be directed at better understanding and improving diversity in terms of performance-relevant characteristics, circumventing the need for detailed patient-level labels.

Acknowledgments. This project was part-funded by the EPSRC CDT in Medical Imaging at King's College London and Imperial College London (EP/S022104/1).

Disclosure of Interests. The authors have no competing interests to declare that are relevant to the content of this article.

References

1. Antonelli, M., et al.: The medical segmentation decathlon. Nat. Commun. **13**(1), 4128 (2022)
2. Castro, D.C., Walker, I., Glocker, B.: Causality matters in medical imaging. Nat. Commun. **11**(1), 3673 (2020)
3. Chen, C., et al.: Enhancing MR image segmentation with realistic adversarial data augmentation. Med. Image Anal. **82** (2022)
4. Chouker, A., et al.: Estimation of liver size for liver transplantation: the impact of age and gender. Liver Transpl. **10**(5), 678–685 (2004)
5. Gava, A., Freitas, F., Meyrelles, S., Silva, I., Graceli, J.: Gender-dependent effects of aging on the kidney. Braz. J. Med. Biol. Res. **44**, 905–913 (2011)
6. Gichoya, J.W., et al.: AI recognition of patient race in medical imaging: a modelling study. Lancet Digit. Health **4**(6), e406–e414 (2022)
7. Guan, H., Liu, M.: Domain adaptation for medical image analysis: a survey. IEEE Trans. Biomed. Eng. **69**(3), 1173–1185 (2021)
8. Hatamizadeh, A., et al.: UNETR: transformers for 3D medical image segmentation. In: Proceedings of the IEEE/CVF Winter Conference on Applications of Computer Vision, pp. 574–584 (2022)
9. Hesamian, M.H., Jia, W., He, X., Kennedy, P.: Deep learning techniques for medical image segmentation: achievements and challenges. J. Digit. Imaging **32**, 582–596 (2019)
10. Ioannou, S., Chockler, H., Hammers, A., King, A.P., Initiative, A.D.N.: A study of demographic bias in CNN-based brain MR segmentation. In: Abdulkadir, A., et al. (eds.) MLCN 2022. LNCS, vol. 13596, pp. 13–22. Springer, Cham (2022). https://doi.org/10.1007/978-3-031-17899-3_2
11. Isensee, F., Jaeger, P.F., Kohl, S.A.A., Petersen, J., Maier-Hein, K.H.: nnU-Net: a self-configuring method for deep learning-based biomedical image segmentation. Nat. Methods **18**(2), 203–211 (2021)
12. Isensee, F., Petersen, J., Kohl, S.A., Jäger, P.F., Maier-Hein, K.H.: nnU-Net: breaking the spell on successful medical image segmentation. arXiv preprint arXiv:1904.08128, vol. 1, no. 1-8, p. 2 (2019)
13. Ji, Y., et al.: Amos: a large-scale abdominal multi-organ benchmark for versatile medical image segmentation. In: Advances in Neural Information Processing Systems, vol. 35, pp. 36722–36732 (2022)

14. Kipp, J.P., Olesen, S.S., Mark, E.B., Frederiksen, L.C., Drewes, A.M., Frøkjær, J.B.: Normal pancreatic volume in adults is influenced by visceral fat, vertebral body width and age. Abdom. Radiol. **44**, 958–966 (2019)
15. Koh, P.W., et al.: Wilds: a benchmark of in-the-wild distribution shifts. In: International Conference on Machine Learning, pp. 5637–5664. PMLR (2021)
16. Kreel, L., Sandin, B.: Changes in pancreatic morphology associated with aging. Gut **14**(12), 962–970 (1973)
17. Larrazabal, A.J., Nieto, N., Peterson, V., Milone, D.H., Ferrante, E.: Gender imbalance in medical imaging datasets produces biased classifiers for computer-aided diagnosis. Proc. Natl. Acad. Sci. **117**(23), 12592–12594 (2020)
18. Lee, T., Puyol-Antón, E., Ruijsink, B., Shi, M., King, A.P.: A systematic study of race and sex bias in CNN-based cardiac MR segmentation. In: Camara, O., et al. (eds.) STACOM 2022. LNCS, vol. 13593, pp. 233–244. Springer, Cham (2022). https://doi.org/10.1007/978-3-031-23443-9_22
19. Li, W., Yuille, A., Zhou, Z.: How well do supervised models transfer to 3D image segmentation? In: International Conference on Learning Representations (2023)
20. Ma, J.: Cutting-edge 3D medical image segmentation methods in 2020: are happy families all alike? arXiv preprint arXiv:2101.00232 (2021)
21. Marcos, R., Correia-Gomes, C., Miranda, H., Carneiro, F.: Liver gender dimorphism: insights from quantitative morphology. Histol. Histopathol. **30**(12), 1431–1437 (2015)
22. Milletari, F., Navab, N., Ahmadi, S.A.: V-net: fully convolutional neural networks for volumetric medical image segmentation. In: 2016 Fourth International Conference on 3D Vision (3DV), pp. 565–571. IEEE (2016)
23. Petersen, S.E., et al.: Reference ranges for cardiac structure and function using cardiovascular magnetic resonance (CMR) in Caucasians from the UK Biobank population cohort. J. Cardiovasc. Mag. Reson. **19**(1) (2016)
24. Sabolić, I., Asif, A.R., Budach, W.E., Wanke, C., Bahn, A., Burckhardt, G.: Gender differences in kidney function. Pflügers Archiv-Eur. J. Physiol. **455**, 397–429 (2007)
25. Wang, R., Chaudhari, P., Davatzikos, C.: Bias in machine learning models can be significantly mitigated by careful training: evidence from neuroimaging studies. Proc. Natl. Acad. Sci. **120**(6), e2211613120 (2023)
26. Wasserthal, J., et al.: Totalsegmentator: robust segmentation of 104 anatomical structures in CT images. arXiv preprint arXiv:2208.05868 (2022)
27. Yang, Y., Zhang, H., Katabi, D., Ghassemi, M.: Change is hard: a closer look at subpopulation shift. arXiv preprint arXiv:2302.12254 (2023)
28. Zhou, K., Liu, Z., Qiao, Y., Xiang, T., Loy, C.C.: Domain generalization: a survey. IEEE Trans. Pattern Anal. Mach. Intell. **45**(4), 4396–4415 (2022)
29. Zhou, Y., et al.: Multi-contrast computed tomography atlas of healthy pancreas. arXiv preprint arXiv:2306.01853 (2023)

BMFT: Achieving Fairness via Bias-Based Weight Masking Fine-Tuning

Yuyang Xue[1]([✉]), Junyu Yan[1], Raman Dutt[1,2], Fasih Haider[1], Jingshuai Liu[1], Steven McDonagh[1], and Sotirios A. Tsaftaris[1]

[1] School of Engineering, The University of Edinburgh, Edinburgh EH9 3FG, UK
{yuyang.xue,junyu.yan,fasih.haider,jliu11,s.mcdonagh,s.tsaftaris}@ed.ac.uk
[2] School of Informatics, The University of Edinburgh, Edinburgh EH8 9AB, UK
{r.dutt}@sms.ed.ac.uk

Abstract. Developing models with robust group fairness properties is paramount, particularly in ethically sensitive domains such as medical diagnosis. Recent approaches to achieving fairness in machine learning require a substantial amount of training data and depend on model retraining, which may not be practical in real-world scenarios. To mitigate these challenges, we propose Bias-based Weight Masking Fine-Tuning (BMFT), a novel *post-processing* method that enhances the fairness of a trained model in significantly fewer epochs without requiring access to the original training data. BMFT produces a mask over model parameters, which efficiently identifies the weights contributing the most towards biased predictions. Furthermore, we propose a two-step debiasing strategy, wherein the feature extractor undergoes initial fine-tuning on the identified bias-influenced weights, succeeded by a fine-tuning phase on a reinitialised classification layer to uphold discriminative performance. Extensive experiments across four dermatological datasets and two sensitive attributes demonstrate that BMFT outperforms existing state-of-the-art (SOTA) techniques in both diagnostic accuracy and fairness metrics. Our findings underscore the efficacy and robustness of BMFT in advancing fairness across various out-of-distribution (OOD) settings. Our code is available at: https://github.com/vios-s/BMFT.

Keywords: Algorithm Fairness · Bias Identification · Bias Removal

1 Introduction

Machine learning is known to exhibit biases from various sources, such as human judgement, inherent algorithmic predispositions, and representation bias [22].

Y. Xue and J. Yan—The two authors share the equal contribution.

Supplementary Information The online version contains supplementary material available at https://doi.org/10.1007/978-3-031-72787-0_10.

The latter emerges when data from minority ethnicities, genders, and age groups are under-represented, leading to unfair predictions by machine learning systems. This can result in serious issues, including direct or indirect discrimination [24]. It is important to ensure an equitable and responsible application of AI to not undermine the trustworthiness and acceptance of AI solutions by end users [32].

Xu et al. [34] identified three key strategies to mitigate bias in medical imaging: pre-processing, e.g. sample re-distribution [27]; in-processing, including learning of disentangled representations [2,19]; and *post-processing*, involving weight pruning [21,33]. While pre- and in-processing are effective in producing fair predictions, a challenge remains: *Access to original medical training data, post-deployment in the real-world, is often restricted.* Mao et al. [20] showed that empirical risk minimisation (ERM) on imbalanced data can effectively capture features for classification within the representation domain. This suggests that post-processing can overcome bias with only a small amount of external data, e.g. by fine-tuning or pruning weights, thus eliminating the need for complete model retraining on the large-scale original training dataset.

Weights learnt with ERM encode two types of features: *core* features that contribute to the target task and *bias* features that have captured spurious correlations or irrelevant information in the data and could therefore exhibit bias during decision-making. It is well understood that these two distinct types of features are entangled in a complex manner [17,35]. Current weight pruning strategies [15,33] aim to identify and remove weights that contribute to bias, based on certain heuristics. However, this can cause substantial performance degradation mainly due to the complex entanglement between *core* features and bias features, especially when a model is trained from random initialisation [30]. It is well known that fine-tuning from a pre-trained initialisation leads to faster convergence and better performance [28]. However, if fine-tuning is indiscriminately applied to all weights in pursuit of fairness, there is the potential of updating *core* features which have previously captured essential discriminative information, therefore risking predictive performance power, even if model fairness gets improved. Moreover, fine-tuning all model parameters with limited data is prone to overfitting [8]. A potential solution is to fine-tune a subset of model weights to lessen the effect of biased feature on the decision boundary [17].

We propose Bias-based Weight Masking Fine-Tuning (BMFT), a fast, retrain-free, post-processing debiasing approach designed to enhance fairness while preserving performance, which utilises a small external dataset and eliminates the need for original training data. Different from other mask-based methods that implement iterative re-training on the original data [7], *BMFT* firstly generates a parameter update mask which highlights the model parameters contributing to bias. Then, a two-step process fine-tunes the feature extractor on mask-selected weights for feature debiasing, which is followed by classification layer fine-tuning to efficiently integrate core features and improve classification performance. This strategy can lead to remarkable performance improvements in terms of prediction fairness and accuracy within significantly fewer training steps.

Our **contributions** can be summarised as follows: **(1)** We propose a fast mask generation technique to filter weights contributing to bias with a small external dataset and without needing the original training data. **(2)** We introduce a two-step fine-tuning strategy that is model-agnostic and contributes to bias mitigation without sacrificing the model's performance. **(3)** We conduct extensive experiments across different datasets and sensitive attributes to reveal the superiority of our proposed approach compared to SOTAs.

2   Method

Bias in model prediction typically originates from two key sources: (1) the *entanglement* between noisy, harmful (spurious or irrelevant) features and, useful, *core* features within the feature extractor; and (2) the *incorrect composition* of representation in the classification layer, leading to the core features information loss [20]. **The main idea** of *BMFT* (Fig. 1) is to start with debiasing the feature extractor, by fine-tuning only the targeted weights that are disproportionately instigating bias. Subsequently, as the first process may affect how the core features are integrated by the classification layer for prediction, *BMFT* then fine-tunes a reinitialised classification layer.

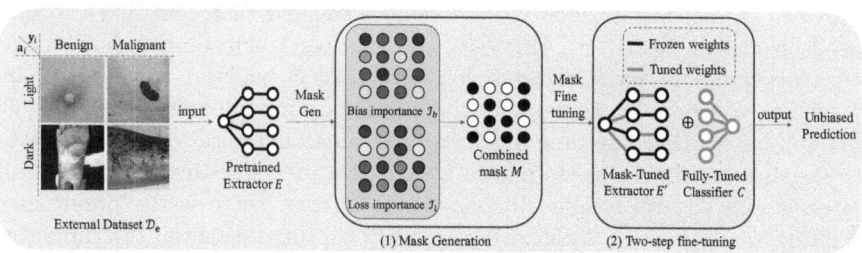

Fig. 1. *BMFT* is a masked-based fine-tuning post-processing approach. (1) A weight mask is generated by calculating weight importance to the bias and loss functions. (2) The masked weights of the feature extractor $E(\cdot)$ are fine-tuned to debias, followed by fine-tuning a reinitialised classification layer $C(\cdot)$ to maintain model performance.

2.1   Preliminaries

Consider an external dataset of N subjects, $\mathcal{D}_e = \{\mathbf{x}_i, \mathbf{y}_i, \mathbf{a}_i\}_{i=1}^N$, where $\mathbf{x}_i$, $\mathbf{y}_i$, and $\mathbf{a}_i$ are the input image, the corresponding label, and the sensitive attribute, respectively. For simplicity, we assume binary targets and attributes (i.e. $\mathbf{y}_i, \mathbf{a}_i \in \{0, 1\}$). We will assume a compositional construction of a biased model $f_\theta(\cdot)$, consisting of a feature extractor $E(\cdot)$ and a classification head $C(\cdot)$, pre-trained by the original training data $\mathcal{D}_{\mathrm{org}}$. Our goal is to debias the model $f_\theta(\cdot)$ w.r.t. fairness metrics such as Demographic Parity (DP) [9] or Equalised

Odds (EOdds) [13] by fine-tuning with significantly fewer epochs, using $\mathcal{D}_e$; and further enhance the generalisability of the model on various OOD test datasets $\mathcal{D}_t$, which share the same sensitive attribute $\mathbf{a}_i$. To avoid attribute bias from the *external dataset* $\mathcal{D}_e$, we sample a *group-balanced subset* $\mathcal{D}_r$ from $\mathcal{D}_e$. This ensures that each attribute group $\mathbf{a}_i$ has the same amount of data. The proportion of positive to negative labelled images is adjusted to be the same in each $\mathbf{a}_i$ group, which prevents bias due to different label distributions. Implementing these two strategies prepares the model for debiasing by providing equal representation across attribute groups and equal ratios of label types. Choosing proportionally over numerical equality in $\mathcal{D}_r$ helps when there are few positive or negative labels in $\mathcal{D}_e$, thus avoiding a reduction in the overall sample size.

2.2   Bias-Importance-Based Mask Generation

To identify the effective parameters for updating, we evaluated the importance of each parameter with respect to the designed bias influence function and loss function. The top K parameters, that are highly correlated with bias and yet provide only minimal contribution to prediction, were selected as the masking weights. We adopt weighted Binary Cross Entropy (WBCE) as loss function since the dermoscopic dataset contains label imbalance. Together with the bias influence function $\mathcal{B}$, the differentiable proxy of EOdds [13], is given by:

$$\mathcal{L}_{\text{WBCE}} = \sum_i \left(-\frac{N_n}{N_n + N_p} \mathbf{y}_i \log(p_i) - \frac{N_p}{N_n + N_p} (1 - \mathbf{y}_i) \log(1 - p_i) \right), \quad (1)$$

$$\mathcal{B}(f_\theta, \mathbf{x}_i, \mathbf{y}_i, \mathbf{a}_i) = \left| \frac{\sum_{i=1}^N f_\theta(\mathbf{x}_i)(1 - \mathbf{a}_i)\mathbf{y}_i}{\sum_{i=1}^N (1 - \mathbf{a}_i)\mathbf{y}_i} - \frac{\sum_{i=1}^N f_\theta(\mathbf{x}_i)\mathbf{a}_i\mathbf{y}_i}{\sum_{i=1}^N \mathbf{a}_i\mathbf{y}_i} \right|, \quad (2)$$

where p_i is the prediction probability, N_p is the number of images with positive label, N_n the number of images with negative label. For a pre-trained model $f_\theta(\cdot)$ with effectively learnt parameters θ^*, the sensitivity for each parameter θ can be obtained by the second-order derivative of loss close to the minimum [10]. This value can be interpreted as the importance of the parameter for the prediction. The diagonal of the Fisher Information Matrix (FIM) is equivalent to the second-order derivative of the loss [26], which provides an efficient computation through first-order derivatives. The first-order expression for the FIM is provided by:

$$\mathcal{I}(f_\theta, \mathcal{D}_e) = \mathbb{E}\left[\left(\frac{\partial \log f_\theta(\mathcal{D}_e|\theta)}{\partial \theta} \right) \left(\frac{\partial \log f_\theta(\mathcal{D}_e|\theta)}{\partial \theta} \right)^\top \Big|_\theta \right]. \quad (3)$$

We employ Eq. 3 to quantify the importance of each weight that contributes to the prediction. The importance of each weight for the bias can be computed using the second derivative of the bias influence function (that is, Eq. 2). Noting that the bias influence function is a linear function of the prediction, and thus replaces the loss function with the bias influence function during backpropagation, the importance of a parameter w.r.t. the bias can also be derived by Eq. 3. We denote the weight mask as M_i, where i is the weight index.

To ensure that the weight importance for bias or prediction is not skewed in favour of a particular subgroup, in practice, we use the group-balanced dataset $\mathcal{D}_r$ to calculate the importance. To select top $K\%$ weights in the mask M_i, we use the bias importance $\mathcal{I}_{i,b}$ and the loss importance $\mathcal{I}_{i,l}$ of each parameter as:

$$M_i = \begin{cases} 1, & \text{if } \mathcal{I}_{i,b}/\mathcal{I}_{i,l} > \alpha \\ 0, & \text{otherwise} \end{cases}, \tag{4}$$

where $\mathcal{I}_{i,b}$ and $\mathcal{I}_{i,l}$ denote the importance of each parameter in relation to bias and loss, respectively. The hyperparameter α distinguishes the significance of the weight to bias-induced features as opposed to core features, calculated by using the weight selection ratio K, defined as the top K-th percentile of the ratios of importance of the parameters. An illustrative example of a mask is provided in the supplementary material.

2.3 Impair-Repair: Two-Step Fine-Tuning

To disentangle core features from biased features, whilst preserving predictive capability, a two-stage fine-tuning process is structured in an "impair-repair" manner. Firstly, we fine-tune the selected masked parameters within the feature extractor $E(\cdot)$ using the group-balanced dataset $\mathcal{D}_r$, guided by an objective function (Eq. 2). Next, we fine-tune the entire reinitialised classification layer $C(\cdot)$. The weights identified by the mask in the feature extractor are closely linked to attribute bias and play a smaller role in predicting the class. This relationship helps us remove the unfavourable influence of bias contained in the model, and yet crucially also retain information embedded in core features.

We highlight that the pre-trained model has the capability to sufficiently extract core features. Further, eliminating the influence of (only) bias-effected weights suggests moving the model decision boundary away from bias features. Moreover, fine-tuning is prone to overfitting on a small dataset. These observations suggest that small epochs are sufficient to realise the proposed debiasing fine-tuning strategy.

An unbiased model, trained with a balanced dataset, will have the majority of its classification layer weights being zero, due to minimisation of irrelevant feature impact on class prediction [17]. To this end, we first reinitialise the weights and biases in the classification layer $C(\cdot)$ to zero to enable fast convergence to the optimal unbiased state and then fine-tune the layer on $\mathcal{D}_r$ using a loss which combines the WBCE loss $\mathcal{L}_{\text{WBCE}}$ and fairness constraints (i.e. Eq. 2), as:

$$\mathcal{L} = \mathcal{L}_{\text{WBCE}}(f_\theta) + \beta \mathcal{B}(f_\theta), \tag{5}$$

where the fairness constraint $\mathcal{B}(f_\theta)$ modulates the emphasis of the features, and the hyperparameter β establishes an efficiently identified core trade-off between class prediction accuracy and fairness. The improved feature extractor retains core features without being influenced by attribute bias, allowing the model to efficiently identify and recover its predictive performance in limited epochs. To balance bias reduction with efficiency, an equal number of epochs is used for both the "impair" and "repair" stages. Our empirical findings indicate that only 10% of the total original training epochs are adequate for successful fine-tuning.

3   Experiments and Results

3.1   Dataset and Data-Processing

ISIC Challenge Training Dataset. The International Skin Imaging Collaboration (ISIC) challenge dataset is a collection of dermoscopic images of melanoma classification, complete with diagnostic labels and metadata. A combination of the 2017 [4], 2019 [5,31] and 2020 [29] ISIC challenge data is used as the training dataset (46,938 images), while the 2018 [3] data is reserved as the *external dataset* $\mathcal{D}_e$ (9,925 images, excluding any overlap with the training data). We select 4,024 images for skin tone and 9,084 for gender from $\mathcal{D}_e$ to build *group-balanced subset* $\mathcal{D}_r$. We maintain the same skin tone annotation as [1] for training data. All images are pre-processed with centre-cropping and resizing to size 256×256. We consider two available sensitive attributes (skin tone and gender).

Test Datasets. Our study employs four different OOD datasets with skin images for melanoma detection. Fitzpatrick-17k [11] contains 16,577 images across six skin tone levels, which we group into light (1-3) and dark (4-6) categories. The PAD-UFES-20 [23] includes 2,298 images, with detailed metadata such as gender, ethnicity, and skin tone. Interactive Atlas of Dermoscopy (Atlas) [18] provides 1,011 images with gender, and DDI [6] offers 656 images with age and skin information. These datasets expand diversity, aiding in evaluating our model's generalisability across attributions. Details are in the supplementary material.

3.2   Implementation

We conducted experiments in PyTorch [25] using NVIDIA A100 40GB GPUs. We adopt different variants of ImageNet pretrained ResNet architecture (ResNet-18, 34, 50, and 101) [14]. The learning rates for impair and repair processes were set at 0.001 and 0.003 respectively, using Adam optimiser [16], with a batch size of 64. We perform a hyperparameters sweep and selected $K = 50$, and β was selected as 0.02 for the ResNet50 model (details on hyperparameter tuning are shown in supplementary materials). To address class-imbalance, we adopt the WBCE loss function $\mathcal{L}_{\text{WBCE}}$. Furthermore, we use accuracy (ACC), area under the receiver operating characteristic curve (AUC) as primary performance metrics, and equalised odds (EOdds [13]) as fairness metric, as done previously [7,12,33]. Our post fine-tuning method requires just an additional five epochs, which is 10% of the initial 50-epoch training duration. Code will be released soon.

3.3   Results

We compare with baselines and SOTA models which we describe below. We pre-trained the ResNet model on the training dataset $\mathcal{D}_{org}$ as a *Baseline*. Performance is evaluated using established metrics. Full fine-tuning (*Full FT*), a widely adopted fine-tuning practice, adjusts all weights on the external dataset $\mathcal{D}_e$ with the loss function specified in Eq. 5. *FairPrune* [33] improves fairness by pruning parameters and prioritises subgroup accuracy; however, the optimal hyperparameter configuration is specific to model and dataset. *LLFT* [20] fine-tunes only the last layer of a deep classification model to promote fairness. Similarly, *Diff-Bias* [21] fine-tunes on an external dataset, using a bias-aware loss function to steer network optimisation. We highlight that a good fairness score does not constitute the sole criterion for success and that a model should also afford accurate classification behaviour, represented here by high ACC and AUC. The results presented here correspond to the ResNet-50 model and analogous trends were observed across the investigated models, with further details present in the supplementary materials.

Table 1. Comparison of debiasing methods on skin tone and gender attributes, for a ResNet-50 model. We re-implemented SOTA methods (denoted with *) or used publicly available code (†). *FairPrune* had difficulty in classifying the Atlas dataset.

Methods	Skin Tone									Gender					
	Fizpatric17k			PAD-UFES-20			DDI			Atlas			PAD-UFES-20		
	ACC↑	AUC↑	EOdds↓	ACC↑	AUC↑	EOdds↓	ACC↑	AUC↑	EOdds↓	ACC↑	AUC↑	EOdds↓	ACC↑	AUC↑	EOdds↓
Baseline	0.719	0.521	0.1745	0.858	0.506	0.0412	0.744	0.646	0.1288	0.574	0.614	0.0348	0.858	0.506	0.0167
Full FT	0.703	0.522	0.1510	0.725	0.501	**0.0321**	0.727	0.618	0.2059	0.690	0.655	0.0303	0.725	0.501	0.0730
FairPrune [33]*	0.763	0.547	0.0120	0.698	0.492	0.1237	0.737	0.536	**0.0024**	\	\	\	0.634	0.499	0.1456
LLFT [20]†	0.579	0.513	0.1446	0.694	0.492	0.0726	0.646	0.561	0.1197	0.724	0.589	0.0042	**0.964**	0.482	**0.0007**
Diff-Bias [21]†	0.716	0.507	0.1144	0.884	0.501	0.0599	0.737	0.629	0.0741	**0.769**	0.687	0.0383	0.878	0.504	0.0461
BMFT (Ours)	**0.865**	**0.637**	**0.0055**	**0.902**	**0.507**	0.0657	**0.759**	**0.736**	0.0754	0.752	**0.876**	**0.0037**	0.893	**0.512**	0.0365

Fine-Tuning Can Help Achieve Fairness. We start by evaluating the behaviour of fine-tuning methods. In terms of fairness, we observed significant improvements both in AUC and EOdds across the majority of test datasets when employing fine-tuning-based methods (Table 1). This indicates that fine-tuning can improve the accuracy gaps between two attribute groups and improve fairness metrics. Fine-tuning with fairness constraints, i.e. *LLFT* and *Diff-Bias*, exhibit lower EOdds compared to using a WBCE loss, i.e. *Full FT*. This illustrates that incorporating the bias influence function into the loss has measurable benefits.

Pruning Is Not Always the Best Option. Model pruning, a popular technique for model simplification towards reducing bias, does not always lead to fairer outcomes, c.f. fine-tuning methods. Our results demonstrate that pruning achieves low AUC across most test datasets, even under-performing the baseline. For example, the performance of *FairPrune* drops drastically under the Atlas dataset, as detailed in Table 1, regardless of hyperparameter values. It is "easier" to obtain good fairness by finding less useful features for classification. This leads to poor classification results in sensitive groups. We conjecture that such performance declines are due to pruning which removes neurons containing *both* bias and core features that are important for classification, therefore causing undesirable information loss. Balancing between fairness and classification performance through pruning requires extensive iterations and is heavily dependent on hyperparameters and datasets. Instead, our method considers both classification capability and fairness to force the selection of core features.

Masked Fine-Tuning Utility. Traditional fine-tuning techniques struggle to achieve a balance between classification performance and fairness. *Full FT* and *Diff-Bias* improve all metrics only in the Atlas dataset for the gender attribute (see Table 1). This illustrates that fine-tuning all layers guides model behaviour towards the distribution of the external dataset, therefore necessitating similar distributions between external and test datasets. Moreover, this may affect weights that are useful for prediction in lieu of fairness, leading to degradation in classification performance. *LLFT* shows improvement in EOdds, but performs poorly in terms of AUC (0.589 in Atlas), showing that fine-tuning the last layer from scratch requires all core features that are well captured and isolated from the bias features; otherwise, fairness and prediction irrelevant features may be selected by the classification layer for achieving fairness. The proposed method, *BMFT*, manages to find an ideal balance and exhibits the best performance in terms of ACC, AUC and EOdds on most test datasets.

Different Bias, Different Difficulty. Comparing *Baseline* results for gender and skin tone attributes in the PAD-UFES-20 dataset (see Table 1), the pre-trained model demonstrates less inherent bias for gender, indicated by a lower EOdds. Furthermore, Table 1 shows that *BMFT* achieves more prominent EOdds and AUC improvements for the gender attribute c.f. the skin tone. This finding suggests that attributes with less inherent bias in the pre-trained model are easier for bias mitigation strategies to rectify. Moreover, the presence of less inherent bias suggests that core and biased features are less entangled, which simplifies the task of maintaining model performance during mitigation process.

Mask: Better Than Manual, Superior to Random. We compare our mask with random masking, manual shallow convolutional (Top) masking and batch normalisation (BN) masking in Table 2. The latter two options are motivated by analysis of the mask distribution, where we saw that mask-selected weights exist mainly in shallow convolutional layers and BN. Compared to random, the results align with our hypothesis that shallow convolutional layers and BN layers contribute to spurious correlation. Our method outperforms the baselines by ~0.2

Table 2. Comparison of the random mask, manual masks, and our proposed mask generation method on both skin and gender attributes, using ResNet-50 backbone.

Methods	DDI (Skin Tone)			Atlas (Gender)		
	ACC↑	AUC↑	EOdds↓	ACC↑	AUC↑	EOdds↓
Random	0.733	0.581	0.1194	0.741	0.629	0.0105
Manual Top Layers	0.749	0.670	0.0948	0.753	0.644	0.0169
Manual BN Layers	0.741	0.635	0.1116	**0.757**	0.661	0.0052
BMFT (Ours)	**0.759**	**0.736**	**0.0754**	0.752	**0.876**	**0.0037**

in AUC, illustrating the effectiveness of our fast mask generation process. With masking, spurious-related weights can be fine-tuned, and core-feature-related weights will remain intact, leading to improved fairness and prediction efficacy.

4   Conclusion

Our study advances bias mitigation in discriminative models for dermatological disease. The proposed *BMFT* distinguishes the core features from biased features, enhancing fairness without sacrificing classification performance. This two-step impair-repair fine-tuning approach can effectively reduce bias within minimal epochs, with only 10% of the original training effort, providing an efficient solution when computing resources or data access are limited. Our method is agnostic to the choice of pre-trained models. Our future work will explore a broader range of tasks and multiple attribute-label pairs. The challenge of obtaining diverse, well-labelled public datasets with comprehensive meta-data or sensitive attributes remains a limitation on our work.

Acknowledgments. Y. Xue and J. Yan thank additional financial support from the School of Engineering, the University of Edinburgh. J. Yan also thanks the support of Advanced Care Research Center. S.A. Tsaftaris also acknowledges the support of Canon Medical and the Royal Academy of Engineering and the Research Chairs and Senior Research Fellowships scheme (grant RCSRF1819\8\25), of the UK's Engineering and Physical Sciences Research Council (EPSRC) (grant EP/X017680/1) and the National Institutes of Health (NIH) grant 7R01HL148788-03. We thank Dr. Edward Moroshko for his help and support.

Disclosure of Interests. The authors have no competing interests to declare that are relevant to the content of this article.

References

1. Bevan, P., Atapour-Abarghouei, A.: Skin deep unlearning: artefact and instrument debiasing in the context of melanoma classification. arXiv:2109.09818 (2021)
2. Bissoto, A., Valle, E., Avila, S.: Debiasing skin lesion datasets and models? Not so fast. In: Proceedings of the IEEE/CVF Conference on Computer Vision and Pattern Recognition Workshops, pp. 740–741 (2020)

3. Codella, N., et al.: Skin lesion analysis toward melanoma detection 2018: a challenge hosted by the international skin imaging collaboration (ISIC). arXiv preprint arXiv:1902.03368 (2019)
4. Codella, N.C., et al.: Skin lesion analysis toward melanoma detection: a challenge at the 2017 international symposium on biomedical imaging (ISBI), hosted by the international skin imaging collaboration (ISIC). In: 2018 IEEE 15th International Symposium on Biomedical Imaging (ISBI 2018), pp. 168–172. IEEE (2018)
5. Combalia, M., et al.: Bcn20000: dermoscopic lesions in the wild. arXiv preprint arXiv:1908.02288 (2019)
6. Daneshjou, R., et al.: Disparities in dermatology AI performance on a diverse, curated clinical image set. Sci. Adv. 8(31), 6147 (2022)
7. Dutt, R., Bohdal, O., Tsaftaris, S.A., Hospedales, T.: FairTune: optimizing parameter efficient fine tuning for fairness in medical image analysis. arXiv preprint arXiv:2310.05055 (2023)
8. Dutt, R., Ericsson, L., Sanchez, P., Tsaftaris, S.A., Hospedales, T.: Parameter-efficient fine-tuning for medical image analysis: the missed opportunity. arXiv preprint arXiv:2305.08252 (2023)
9. Dwork, C., Hardt, M., Pitassi, T., Reingold, O., Zemel, R.: Fairness through awareness. In: Proceedings of the 3rd Innovations in Theoretical Computer Science Conference, pp. 214–226 (2012)
10. Foster, J., Schoepf, S., Brintrup, A.: Fast machine unlearning without retraining through selective synaptic dampening. arXiv preprint arXiv:2308.07707 (2023)
11. Groh, M., et al.: Evaluating deep neural networks trained on clinical images in dermatology with the fitzpatrick 17k dataset. In: Proceedings of the IEEE/CVF Conference on Computer Vision and Pattern Recognition, pp. 1820–1828 (2021)
12. Han, S.S., Kim, M.S., Lim, W., Park, G.H., Park, I., Chang, S.E.: Classification of the clinical images for benign and malignant cutaneous tumors using a deep learning algorithm. J. Investig. Dermatol. 138(7), 1529–1538 (2018)
13. Hardt, M., Price, E., Srebro, N.: Equality of opportunity in supervised learning. In: Advances in Neural Information Processing Systems, vol. 29 (2016)
14. He, K., Zhang, X., Ren, S., Sun, J.: Deep residual learning for image recognition. In: Proceedings of the IEEE Conference on Computer Vision and Pattern Recognition, pp. 770–778 (2016)
15. Huang, Y.Y., Chiuwanara, V., Lin, C.H., Kuo, P.C.: Mitigating bias in MRI-based alzheimer's disease classifiers through pruning of deep neural networks. In: Wesarg, S., et al. (eds.) Workshop on Clinical Image-Based Procedures, pp. 163–171. Springer, Cham (2023). https://doi.org/10.1007/978-3-031-45249-9_16
16. Kingma, D.P., Ba, J.: Adam: a method for stochastic optimization. arXiv preprint arXiv:1412.6980 (2014)
17. Le, P.Q., Schlötterer, J., Seifert, C.: Is last layer re-training truly sufficient for robustness to spurious correlations? arXiv preprint arXiv:2308.00473 (2023)
18. Lio, P.A., Nghiem, P.: Interactive atlas of dermoscopy. J. Am. Acad. Dermatol. 50(5), 807–808 (2004)
19. Liu, X., Thermos, S., Chartsias, A., O'Neil, A., Tsaftaris, S.A.: Disentangled representations for domain-generalized cardiac segmentation. In: Puyol Anton, E., et al. (eds.) STACOM 2020. LNCS, vol. 12592, pp. 187–195. Springer, Cham (2021). https://doi.org/10.1007/978-3-030-68107-4_19
20. Mao, Y., Deng, Z., Yao, H., Kawaguchi, K., Zou, J.: Last-layer fairness fine-tuning is simple and effective for neural networks. arXiv preprint arXiv:2304.03935 (2023)

21. Marcinkevics, R., Ozkan, E., Vogt, J.E.: Debiasing deep chest X-ray classifiers using intra-and post-processing methods. In: Machine Learning for Healthcare Conference, pp. 504–536. PMLR (2022)

22. Mehrabi, N., Morstatter, F., Saxena, N., Galstyan, A.: A survey on bias and fairness in machine learning. arxiv 2019. arXiv preprint arXiv:1908.09635 (2019)

23. Pacheco, A.G., Lima, G.R., Salomao, A.S., Krohling, B., Biral, I.P., de Angelo, G.G., et al.: PAD-UFES-20: a skin lesion dataset composed of patient data and clinical images collected from smartphones. Data Brief **32**, 106221 (2020)

24. Pagano, T.P., Loureiro, R.B., et al.: Bias and unfairness in machine learning models: a systematic review on datasets, tools, fairness metrics, and identification and mitigation methods. Big Data Cogn. Comput. **7**(1), 15 (2023)

25. Paszke, A., et al.: PyTorch: an imperative style, high-performance deep learning library. In: Advances in Neural Information Processing Systems, vol. 32 (2019)

26. Pawitan, Y.: In All Likelihood: Statistical Modelling and Inference using Likelihood. Oxford University Press (2001)

27. Puyol-Antón, E., et al.: Fairness in cardiac MR image analysis: an investigation of bias due to data imbalance in deep learning based segmentation. In: MICCAI 2021. LNCS, vol. 12903, pp. 413–423. Springer, Cham (2021). https://doi.org/10.1007/978-3-030-87199-4_39

28. Raghu, M., et al.: Transfusion: understanding transfer learning for medical imaging. In: Advances in Neural Information Processing Systems, vol. 32 (2019)

29. Rotemberg, V., et al.: A patient-centric dataset of images and metadata for identifying melanomas using clinical context. Sci. Data **8**(1), 34 (2021)

30. Tran, C., Fioretto, F., et al.: Pruning has a disparate impact on model accuracy. Adv. Neural. Inf. Process. Syst. **35**, 17652–17664 (2022)

31. Tschandl, P., Rosendahl, C., Kittler, H.: The HAM10000 dataset, a large collection of multi-source dermatoscopic images of common pigmented skin lesions. Sci. Data **5**(1), 1–9 (2018)

32. Winkler, J.K., et al.: Association between surgical skin markings in dermoscopic images and diagnostic performance of a deep learning convolutional neural network for melanoma recognition. JAMA Dermatol. **155**(10), 1135–1141 (2019)

33. Wu, Y., Zeng, D., Xu, X., Hu, J.: FairPrune: Achieving fairness through pruning for dermatological disease diagnosis. In: Wang, L., Dou, Q., Fletcher, P.T., Speidel, S., Li, S. (eds.) International Conference on Medical Image Computing and Computer-Assisted Intervention, pp. 743–753. Springer, Cham (2022). https://doi.org/10.1007/978-3-031-16431-6_70

34. Xu, Z., Li, J., Yao, Q., Li, H., Zhou, S.K.: Fairness in medical image analysis and healthcare: a literature survey. Authorea Preprints (2023)

35. Ye, H., Zou, J., Zhang, L.: Freeze then train: towards provable representation learning under spurious correlations and feature noise. In: International Conference on Artificial Intelligence and Statistics, pp. 8968–8990. PMLR (2023)

Using Backbone Foundation Model for Evaluating Fairness in Chest Radiography Without Demographic Data

Dilermando Queiroz[1(✉)], André Anjos[2], and Lilian Berton[1]

[1] Universidade Federal de São Paulo, São Paulo, Brazil
dilermando.queiroz@unifesp.br
[2] Idiap Research Institute, Martigny, Switzerland

Abstract. Ensuring consistent performance across diverse populations and incorporating fairness into machine learning models are crucial for advancing medical image diagnostics and promoting equitable healthcare. However, many databases do not provide protected attributes or contain unbalanced representations of demographic groups, complicating the evaluation of model performance across different demographics and the application of bias mitigation techniques that rely on these attributes. This study aims to investigate the effectiveness of using the backbone of Foundation Models as an embedding extractor for creating groups that represent protected attributes, such as gender and age. We propose utilizing these groups in different stages of bias mitigation, including pre-processing, in-processing, and evaluation. Using databases in and out-of-distribution scenarios, it is possible to identify that the method can create groups that represent gender in both databases and reduce in 4.44% the difference between the gender attribute in-distribution and 6.16% in out-of-distribution. However, the model lacks robustness in handling age attributes, underscoring the need for more fundamentally fair and robust Foundation models. These findings suggest a role in promoting fairness assessment in scenarios where we lack knowledge of attributes, contributing to the development of more equitable medical diagnostics.

Keywords: Fairness · Medical Image · Foundation Model

1 Introduction

Recent advancements in medical diagnosis, particularly through Deep Learning (DL) techniques and cloud computing, have the potential to enhance diagnostic accuracy and accessibility. For example, cloud-based DL systems can streamline diagnostics across hospitals, providing crucial tools for medical professionals. However, the rapid proliferation of DL algorithms in healthcare raises ethical concerns about their impact on underrepresented communities [11]. Studies show that Artificial Intelligence (AI) can identify causal structures in data correlated

E. Puyol-Antn et al. (Eds.): FAIMI 2024/EPIMI 2024, LNCS 15198, pp. 109–118, 2025.
https://doi.org/10.1007/978-3-031-72787-0_11

with protected characteristics such as race, gender, age, and ethnicity [7], potentially exacerbating healthcare inequalities by using these correlations to predict health outcomes.

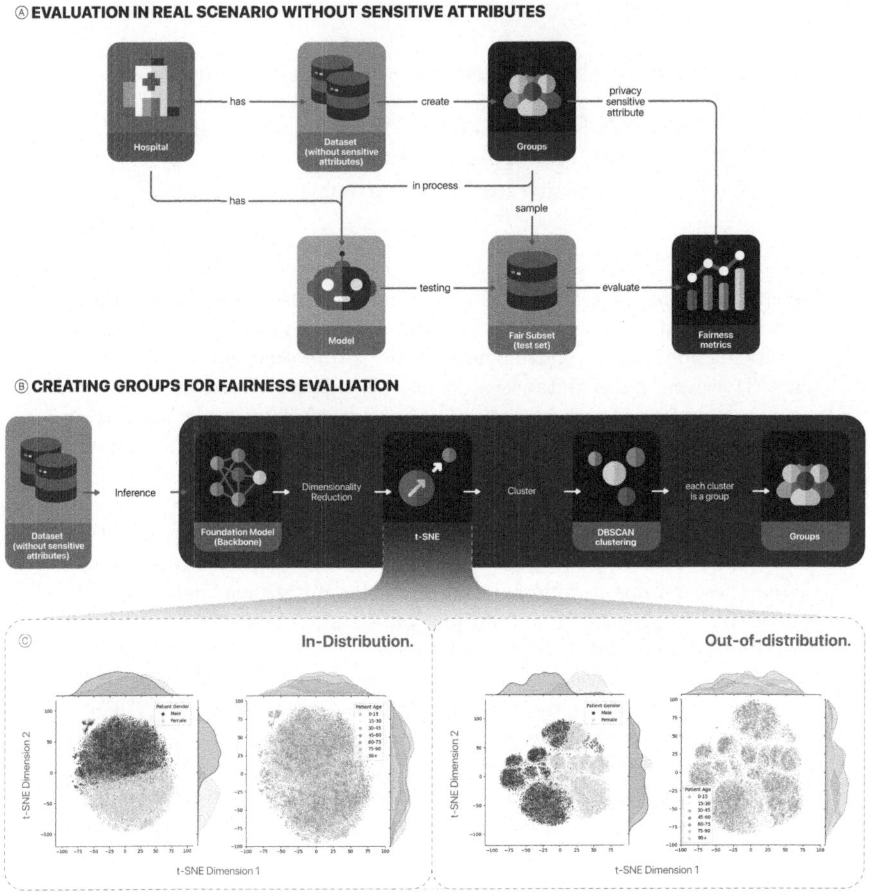

Fig. 1. (a) Overview of the application of groups formed by the proposed method in various contexts such as model processing, subset selection, and metric evaluation. (b) The process begins with a Foundation Model (FM), trained on a large corpus of chest X-ray images, to extract embeddings from a dataset devoid of sensitive attributes. These embeddings are then subjected to dimensionality reduction via t-SNE [10], facilitating clustering in a lower-dimensional space and enhancing computational efficiency. Subsequently, DBSCAN [5] is applied to identify clusters that will be used to form a notion of groups. (c) Visualization of embeddings, which were subsequently reduced to two dimensions using t-SNE. These dimensions are denoted by patient age and gender, spanning across the CheXpert (in-distribution) and NIH (out-of-distribution) databases.

A common issue with medical image data is that many publicly available datasets do not provide demographic information. For example, in chest X-ray datasets, only a few include protected attributes [6], making it challenging to evaluate DL models trained on such data across different demographics and sensitive variables. Moreover, the vast majority of fairness techniques require datasets containing this information [3]. These methods span pre-processing, in-processing, and post-processing stages and are recommended by global health authorities like the World Health Organization to foster equity in healthcare. Therefore, techniques to mitigate unfair treatment from DL models that do not rely on these attributes are essential for the development of the field.

The introduction of new techniques, such as self-supervised learning (SSL), represents a significant leap forward. The use of more images without the need for specific labels using SSL in healthcare allows Foundation Models (FM) to be trained with large quantities of unannotated data, bypassing expensive and tedious labeling processes. An FM is a model trained on extensive and diverse data, typically employing self-supervision at scale, which can then be adapted or fine-tuned for a variety of downstream tasks. Furthermore, FM can help in building more robust models that can be used in a variety of distribution data [2].

The contributions of this work are delineated as follows:

- We leverage the backbone of an FM to construct groups that approximate sensitive attributes, facilitating fairness evaluation methods for datasets without demographic data.
- We propose a comprehensive evaluation and bias mitigation framework tailored for contexts lacking demographic attributes in medical images.
- We show that the FM used in the framework is more robust for gender than for age in demographic data.

2 Related Work

In recent years, there has been an increasing amount of literature on fairness without demographic bias. A notable contribution in this area categorizes methods for achieving fairness without demographic data into four groups: collecting demographic data, implementing additional protections for data collection, utilizing auxiliary datasets and inferring demographic data, and exploring alternatives to traditional group fairness approaches [1]. These methods aim to address the challenges posed by the lack of accurate, complete, or available demographic information, thereby providing a comprehensive framework for researchers, practitioners, and policymakers to navigate the complex landscape of algorithmic fairness. Our method aligns with the category of inferring demographic data. However, these methods exhibit variable performance that can disproportionately impact already marginalized groups [12,13]. We propose using the Foundation Model to extract image embeddings without inferring protected attributes. Due to self-supervised learning, these models are trained solely without labels and can reduce bias [15].

Our proposed method exhibits similarities with the cluster-based balancing approach referred to as fair class balancing [18]. Both methodologies exploit the intrinsic group structure present within the data, identifying naturally occurring homogeneous subgroups characterized by shared feature similarities through clustering in the feature space. By employing these clusters, our method ensures that the training process adequately represents the diversity within the dataset without directly predicting sensitive attributes. We aim to demonstrate that these groups are representative and can be used for fairness evaluation and other applications.

A similar approach incorporates clustering to enhance model fairness and robustness in end-to-end speech recognition systems [16]. While both workflows employ clustering strategies, our method focuses on image data rather than speech data. In their approach, embeddings are used to create clusters, which are utilized during model training to address privacy concerns and improve fairness. Similarly, our method aims to enhance fairness without directly using protected attributes, highlighting the versatility and applicability of clustering techniques across different data domains for promoting fairness and preserving privacy.

3   Materials and Methods

In this section, we describe the methodology used to conduct our research. The real scenario process is illustrated in Fig. 1a, and Fig. 1b shows a detailed flowchart of each step proposed in the work. In the following paragraphs, we detail the method.

Datasets. For the analysis of the framework, two radiology image datasets were employed: CheXpert [8] and ChestX-ray14 (US National Institutes of Health (NIH)) [17]. We chose CheXpert as the in-distribution (ID) dataset, as the base model of the FM was pre-trained using this set. To evaluate the generalization capability of the technique across different datasets, we adopted NIH as an out-of-distribution (OOD) dataset, thereby ensuring the robustness of the approach.

CheXpert contains 224,316 chest radiographs from 65,240 patients, sourced from Stanford Hospital (2002–2017). Only one image per patient was selected, focusing on those with the five most common pathologies (atelectasis, consolidation, pulmonary edema, pleural effusion, and cardiomegaly), resulting in 58,662 images. The NIH dataset includes 112,120 annotated X-rays from 30,805 patients. Similarly, one image per patient was chosen, yielding 30,802 images.

Both datasets provide metadata on age and gender. CheXpert has 55.4% male and 44.6% female patients, while NIH has 53.9% male and 46.1% female. Age distributions differ, with NIH being more unbalanced. Most samples in both datasets are from patients aged 45–65 years, with CheXpert having more older individuals and limited samples from younger individuals. The FM was trained on CheXpert, which lacks pediatric data, whereas NIH includes this age group.

The datasets are the first component of the methodology that as shown in Fig. 1

Extract Embedings. The chosen FM for the method was REMEDIS [2], with input images of 448×448 pixels and three channels, trained from the BiT-M [9] backbone. According to the description, the model underwent initial pre-training on an extensive set of natural images, followed by a second phase of pre-training using self-supervised learning. The specific technique used for pre-training and learning representations is SimCLR [4].

We utilized a REMEDIS pre-trained backbone to extract embeddings from selected images in both datasets. This backbone was employed without further training and used solely for image inference. To address the challenge of visualizing and the computational cost of high-dimensional embeddings, we employed the t-SNE method [10]. This technique reduces the embeddings to two dimensions, making visualization possible.

Clusters. Previous research has established some notion of fairness [3], the three commonly used fairness criteria for binary classification tasks are demographic parity, predictive parity, and equalized odds. These definitions establish a set of (X, Y, A) where X are the samples, Y are the labels, and A are the protected attributes. However, using these metrics is challenging in databases without protected attributes.

To address this, we categorized the data into sets representing specific image characteristics after extracting and reducing the dimensionality of the embeddings. We used DBSCAN [5] to group images by similar characteristics in feature space. This approach allows each image to be assigned to a cluster representing a protected attribute, thus forming the set A using cluster IDs.

The number of clusters directly impacts their size: fewer clusters create larger, more generalized groups, while more clusters result in smaller, more specific groups. We chose to use 15 to 25 clusters to create medium-sized groups. Accordingly, we adjusted the DBSCAN [5] parameters to achieve this range while minimizing unclustered data labeled as -1.

In this approach, we set the minimum number of samples required for each cluster to 120 for the CheXpert dataset and 40 for the NIH dataset. The maximum distance between two samples was set to 4 for CheXpert and 3 for NIH. We determined the number of clusters to be 15 for CheXpert and 22 for NIH, with 14,785 unclustered data points for CheXpert and 2,879 for NIH. To balance the database, we utilized the previously defined clusters and sampled an equal number of samples from each cluster to reach 30% of the original dataset. Since each cluster represents a set of protected attributes, this approach allows us to have a more representative sample of our database.

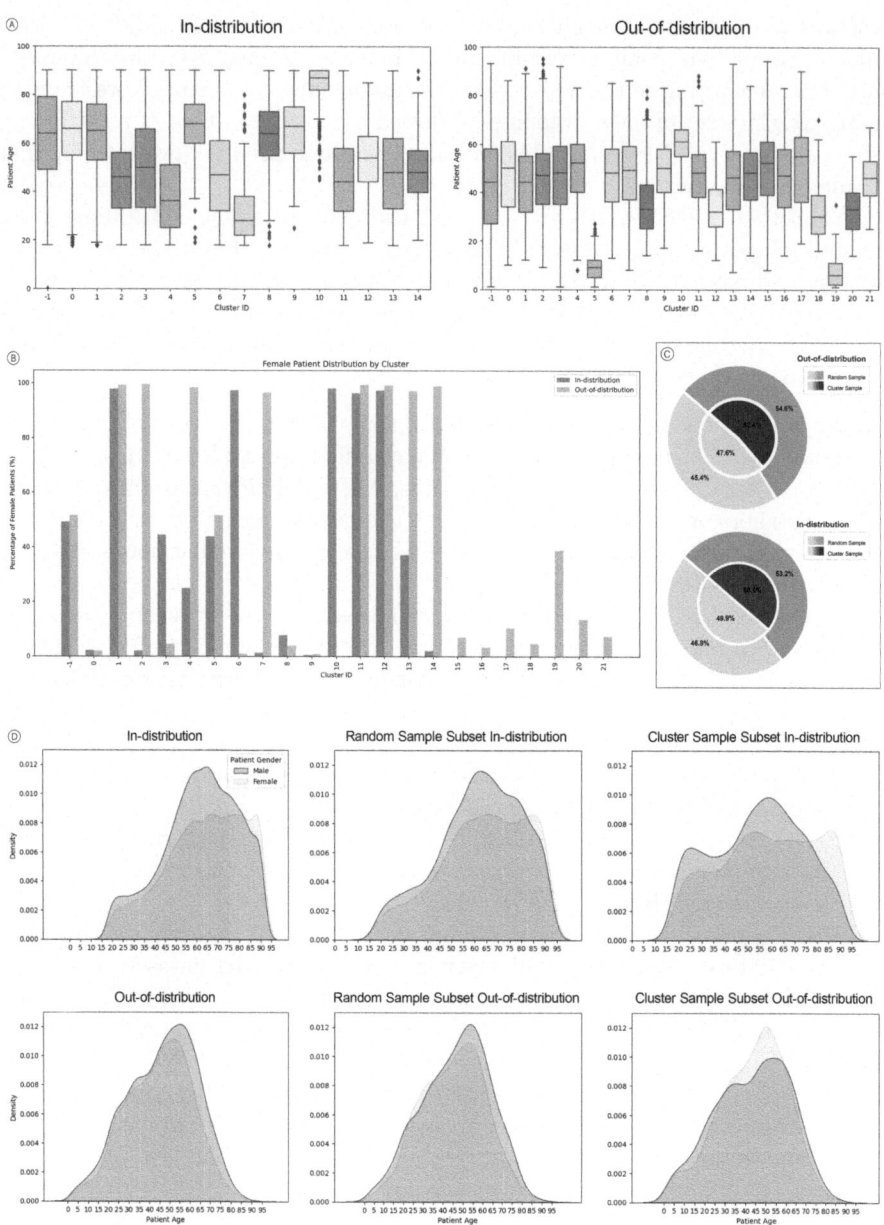

Fig. 2. (a) Age distribution across clusters in the CheXpert (in-distribution) and NIH databases (out-of-distribution). The cluster −1 represents the unclustered data; other numbers are the clusters. **(b)** Gender Distribution Across Clusters for CheXpert and NIH Datasets. **(c)** Gender Distribution after sampling 30% of CheXpert and NIH Datasets. **(d)** Kernel density estimate (KDE) plots illustrate the comparison of age distributions between the CheXpert and NIH subsets, utilizing both random and cluster sampling techniques, with data categorized by patient gender.

4   Results and Discussion

Feature Space Analysis. The first set of analyses examined the proposed method's impact on evaluating fairness in datasets without protected attributes. We analyze the feature spaces of the FM using the metadata about age and gender from both databases, and these attributes are used for visualization and verification of the representative of each generated group. Figure 1c shows the embeddings after dimensional reduction to two dimensions, facilitating the visualization of images extracted by the backbone of the FM. We can visualize strong separation based on the gender attribute in both databases.

However, the distinction based on age is not as clear in both databases. A benchmark study comparing methods for improving fairness using a supervised learning approach [19] shows that the methods yield better results for gender than age across different approaches. As demonstrated in Fig. 1c, self-supervised learning exhibits the same issue. These results suggest that models trained on radiography images generally have more difficulty separating the age attribute in the feature space than the gender attribute. As shown in Fig. 1c in the NIH database, the 0–15 years is the only group clearly defined. The other groups are more mixed, making it more challenging to visualize a separation.

Age and Gender Representation in Clusters. Figure 2a compares the age distribution across clusters for both datasets. Clusters representing more common age groups are more prevalent than those for less common age groups. In the in-distribution dataset, we have clusters represented by different age distributions. However, some clusters share the same distribution. Interestingly, clusters with the same age distribution may correspond to distinctly different groups concerning protected attributes. For instance, clusters 0 and 1 exhibit similar age profiles but differ significantly in gender composition. Cluster 0 contains only 2.20% females, in contrast to cluster 1, which is composed of 97.86% females as shown in Fig. 2b. This gender metadata can be more distinctly segregated, as shown by the clusters for both datasets, where we have clusters with an equal proportion of males and females and clusters composed entirely of one gender.

It is noticeable that within the CheXpert dataset, only one cluster represents individuals aged over 80 years, and similarly, only one cluster for those under 40 years. In contrast, most clusters correspond to more prevalent represented data in the out-of-distribution dataset. Only two groups, clusters 5 and 19, define underrepresented data. Surprisingly, the age range of these clusters is between 0 and 20 years, a demographic not present in the in-distribution dataset, indicating that the model has not encountered this data during training.

These groups must be used with caution because they may contain biases towards certain groups. As they are not exclusively defined by a single protected attribute, there may be small, nested groups that are overlooked in the evaluation process. For instance, in Fig. 2a, within the out-of-distribution scenario, there is an absence of a group representing individuals over the age of 70, likely because they are included within other overlapping groups.

Generating Balanced Subsets. We leveraged the fact that these groups might represent protected variables, thus enabling us to sample an equal number of individuals from each group. With this approach, groups more prevalent in the dataset are combined into the same cluster, reducing the likelihood of these groups being over-sampled. Figure 2d compares two techniques: random sampling and the cluster sampling method proposed in this paper. We utilized the kernel density estimate (KDE) plot to visualize the distribution of the patient age and patient gender in subsets sampled from the original database.

As demonstrated in Fig. 2d, the cluster sampling method yields favorable results for the CheXpert database, which contains in-distribution data. The figure shows a clear trend of a decrease in the majority group regarding age and sex in the cluster sample, resulting in a more balanced sampling compared to the random approach. The random sampling exhibits an imbalance that mirrors the original database regarding age and gender.

A plausible explanation for these observations is that groups with a higher probability of occurrence, such as the age range of 55 to 85 years, are grouped into a single cluster or multiple clusters, as illustrated in Fig. 2a. This confirms that these groups can be used to represent protected attributes and, consequently, be sampled to produce a more balanced subset.

However, the results are insignificant for the NIH database, which contains out-of-distribution data. Figure 2a shows a minor increase in younger age groups and a significant increase in gender distribution. In this case, the random sampling again closely resembles the distribution of the dataset, as expected.

Gender Balance in Sub-datasets. The results regarding the gender proportion in the subsets are presented in Table 2c. In contrast, the cluster sample reduces this disparity, approaching a more balanced distribution of nearly 50%. Specifically for the CheXpert database, the new subset created through cluster sampling exhibits only a 4.72% difference between female and male representation. In comparison, this difference is 9.16% in the random sample, indicating a significant improvement of 4.44% towards gender balance with the cluster sampling approach.

Using random sampling on the NIH dataset, we achieved a subset with a gender difference of 6.4%. However, by employing the technique proposed in the article, we reduced this difference to a mere 0.24%, resulting in an improvement of 6.16%. This method demonstrates robustness for out-of-distribution data in terms of gender attributes. However, it yields slight improvement for the age attribute.

Robustness Across Distribution Shifts. An open question in the literature is whether there are methods that ensure the transfer of fairness across distribution shifts [14]. This study demonstrates the feasibility of maintaining fairness for gender across different distributions using the proposed method. Such an outcome is attributable to using FM as an embedding extractor. However, the method needs to prove more efficient for the age attribute in out-of-distribution.

5   Conclusion

The study demonstrates a novel approach to promoting fairness in medical image diagnostics, especially when demographic data are unavailable. By leveraging the backbone of Foundation Models to create groups representing protected attributes like gender and age, we can apply mitigation techniques across pre-processing, in-processing, and evaluation stages. The results show significant improvements in gender fairness, with a 4.44% and 6.16% reduction in gender attribute disparity for in-distribution and out-of-distribution data, respectively. However, the model faces challenges with age-related attributes, suggesting a need for further development in this area. This research underscores the potential of FMs to advance equitable healthcare diagnostics by providing a framework for fairness evaluation even in the absence of explicit demographic metadata.

Acknowledgments. We thanks Fundação de Amparo à Pesquisa do Estado de São Paulo (FAPESP), grants 21/14725-3 and 23/12493-3, Conselho Nacional de Desenvolvimento Científico e Tecnológico (CNPq), Swiss National Science Foundation (SNSF) under Grant No. 200021E_214653. In addition, we would like to especially thank Lucas Tosta for his assistance in designing the figures.

References

1. Ashurst, C., Weller, A.: Fairness without demographic data: a survey of approaches. In: Proceedings of the 3rd ACM Conference on Equity and Access in Algorithms, Mechanisms, and Optimization, EAAMO 2023, pp. 1–12. Association for Computing Machinery, New York (2023). https://doi.org/10.1145/3617694.3623234. https://dl.acm.org/doi/10.1145/3617694.3623234
2. Azizi, S., et al.: Robust and data-efficient generalization of self-supervised machine learning for diagnostic imaging. Nat. Biomed. Eng. **7**(6), 756–779 (2023). https://doi.org/10.1038/s41551-023-01049-7. https://www.nature.com/articles/s41551-023-01049-7
3. Chen, R.J., et al.: Algorithmic fairness in artificial intelligence for medicine and healthcare. Nat. Biomed. Eng. **7**(6), 719–742 (2023). https://doi.org/10.1038/s41551-023-01056-8. https://www.nature.com/articles/s41551-023-01056-8
4. Chen, T., Kornblith, S., Norouzi, M., Hinton, G.: A Simple Framework for Contrastive Learning of Visual Representations (2020). https://doi.org/10.48550/arXiv.2002.05709. http://arxiv.org/abs/2002.05709. arXiv:2002.05709
5. Ester, M., Kriegel, H.P., Sander, J., Xu, X.: A density-based algorithm for discovering clusters in large spatial databases with noise. In: Proceedings of the Second International Conference on Knowledge Discovery and Data Mining, Portland, Oregon, KDD 1996, pp. 226–231. AAAI Press (1996)
6. Gichoya, J.W., et al.: AI recognition of patient race in medical imaging: a modelling study. Lancet Digit. Health **4**(6), e406–e414 (2022). https://doi.org/10.1016/S2589-7500(22)00063-2. https://www.thelancet.com/journals/landig/article/PIIS2589-7500(22)00063-2/fulltext
7. Glocker, B., Jones, C., Bernhardt, M., Winzeck, S.: Algorithmic encoding of protected characteristics in chest X-ray disease detection models. eBioMedicine **89** (2023). https://doi.org/10.1016/j.ebiom.2023.104467. https://www.thelancet.com/journals/ebiom/article/PIIS2352-3964(23)00032-4/fulltext

8. Johnson, A.E.W., et al.: MIMIC-CXR, a de-identified publicly available database of chest radiographs with free-text reports. Sci. Data **6**(1), 317 (2019). https://doi.org/10.1038/s41597-019-0322-0. https://www.nature.com/articles/s41597-019-0322-0

9. Kolesnikov, A., et al.: Big Transfer (BiT): General Visual Representation Learning (2020). https://doi.org/10.48550/arXiv.1912.11370. http://arxiv.org/abs/1912.11370. arXiv:1912.11370

10. Maaten, L.V.D., Hinton, G.: Visualizing data using t-SNE. J. Mach. Learn. Res. **9**(86), 2579–2605 (2008). http://jmlr.org/papers/v9/vandermaaten08a.html

11. McCradden, M.D., Joshi, S., Mazwi, M., Anderson, J.A.: Ethical limitations of algorithmic fairness solutions in health care machine learning. Lancet Digit. Health **2**(5), e221–e223 (2020). https://doi.org/10.1016/S2589-7500(20)30065-0

12. Scheuerman, M.K., Paul, J.M., Brubaker, J.R.: How computers see gender: an evaluation of gender classification in commercial facial analysis services. Proc. ACM Hum.-Comput. Interact. **3**(CSCW), 1–33 (2019). https://doi.org/10.1145/3359246. https://dl.acm.org/doi/10.1145/3359246

13. Scheuerman, M.K., Wade, K., Lustig, C., Brubaker, J.R.: How we've taught algorithms to see identity: constructing race and gender in image databases for facial analysis. Proc. ACM Hum.-Comput. Interact. **4**(CSCW1), 1–35 (2020). https://doi.org/10.1145/3392866. https://dl.acm.org/doi/10.1145/3392866

14. Schrouff, J., et al.: Diagnosing failures of fairness transfer across distribution shift in real-world medical settings (2023). https://doi.org/10.48550/arXiv.2202.01034. http://arxiv.org/abs/2202.01034. arXiv:2202.01034

15. Vaidya, A., et al.: Demographic bias in misdiagnosis by computational pathology models. Nat. Med. **30**(4), 1174–1190 (2024). https://doi.org/10.1038/s41591-024-02885-z. https://www.nature.com/articles/s41591-024-02885-z

16. Veliche, I.E., Fung, P.: Improving fairness and robustness in end-to-end speech recognition through unsupervised clustering. In: ICASSP 2023 - 2023 IEEE International Conference on Acoustics, Speech and Signal Processing (ICASSP), Rhodes Island, Greece, pp. 1–5. IEEE (2023). https://doi.org/10.1109/ICASSP49357.2023.10096836. https://ieeexplore.ieee.org/document/10096836/

17. Wang, X., Peng, Y., Lu, L., Lu, Z., Bagheri, M., Summers, R.M.: ChestX-Ray8: hospital-scale chest X-ray database and benchmarks on weakly-supervised classification and localization of common thorax diseases. In: 2017 IEEE Conference on Computer Vision and Pattern Recognition (CVPR), pp. 3462–3471 (2017). https://doi.org/10.1109/CVPR.2017.369. https://ieeexplore.ieee.org/document/8099852. ISSN: 1063-6919

18. Yan, S., Kao, H.T., Ferrara, E.: Fair class balancing: enhancing model fairness without observing sensitive attributes. In: Proceedings of the 29th ACM International Conference on Information & Knowledge Management, CIKM 2020, pp. 1715–1724. Association for Computing Machinery, New York (2020). https://doi.org/10.1145/3340531.3411980. https://dl.acm.org/doi/10.1145/3340531.3411980

19. Zong, Y., Yang, Y., Hospedales, T.: MEDFAIR: Benchmarking Fairness for Medical Imaging (2023). https://doi.org/10.48550/arXiv.2210.01725. http://arxiv.org/abs/2210.01725. arXiv:2210.01725

Do Sites Benefit Equally from Distributed Learning in Medical Image Analysis?

Raissa Souza[1,2,3,4(✉)] ⬥, Emma A. M. Stanley[1,2,3,4] ⬥, Richard Camicioli[5], Oury Monchi[1,2,6,7,8], Zahinoor Ismail[2,8,9,10], Matthias Wilms[2,4,11,12] ⬥, and Nils D. Forkert[1,2,4,8] ⬥

[1] Department of Radiology, Cumming School of Medicine, University of Calgary, Calgary, Canada
raissa.souzadeandrad@ucalgary.ca
[2] Hotchkiss Brain Institute, University of Calgary, Calgary, Canada
[3] Biomedical Engineering Graduate Program, University of Calgary, Calgary, Canada
[4] Alberta Children's Hospital Research Institute, University of Calgary, Calgary, Canada
[5] Neuroscience and Mental Health Institute and Department of Medicine (Neurology), University of Alberta, Edmonton, Canada
[6] Department of Radiology, Radio-oncology and Nuclear Medicine, Université de Montréal, Montréal, Canada
[7] Centre de Recherche, Institut Universitaire de Gériatrie de Montréal, Montréal, Canada
[8] Department of Clinical Neurosciences, Cumming School of Medicine, University of Calgary, Calgary, Canada
[9] Department of Psychiatry, University of Calgary, Calgary, Canada
[10] Clinical and Biomedical Sciences, Faculty of Health and Life Sciences, University of Exeter, Exeter, UK
[11] Department of Pediatrics, University of Calgary, Calgary, Canada
[12] Department of Community Health Sciences, University of Calgary, Calgary, Canada

Abstract. Artificial intelligence (AI) has the potential to make medical image analysis more accessible to healthcare institutions worldwide. However, when trained on multi-site datasets, models may excel with data from certain institutions but struggle with others, even when exposed to their training data. This emphasizes the importance of investigating whether all sites benefit from AI models, especially within distributed learning setups. Distributed learning methods allow access to broader and more diverse datasets from multiple sites during training. In this context, the travelling model (TM) paradigm has demonstrated superior performance in limited data scenarios, making it particularly relevant in low-resource settings. This work investigates whether all sites

M. Wilms and N.D. Forkert—Shared last authorship.

Supplementary Information The online version contains supplementary material available at https://doi.org/10.1007/978-3-031-72787-0_12.

can benefit from TM development and identifies the factors influencing performance disparities. Specifically, a Parkinson's disease (PD) database comprising 1,817 neuroimaging datasets from 83 different sites is utilized to investigate how site-specific and participant-specific factors influence the performance of TM in classifying PD. Therefore, we analyze the false positive rate (FPR) and false negative rate (FNR) to identify the characteristics contributing to misdiagnosis. Our findings reveal disparities in benefits across sites, with class imbalance emerging as the major factor influencing FPR and FNR, especially for sites with more PD cases. This research underscores the urgency of a rigorous analysis of a model's behaviour in distributed setups to detect misdiagnosis risks and encourage developers to evaluate and optimize models beyond overall accuracy. Thus, comprehensive analyses of this type can help pave the way for the development of more equitable AI-based medical imaging models.

Keywords: Distributed learning · Travelling model · Site benefit · Health equity

1   Introduction

In the last decade, artificial intelligence (AI) has become a crucial tool for advancing medical image analysis. However, these breakthroughs have predominantly occurred and been implemented in high-income countries (HICs), which possess the resources to collect, store, and curate the imaging databases essential for successful AI development. This disparity may exacerbate healthcare inequities, especially impacting institutions located in remote, under-resourced, and low- and middle-income countries (LMICs) [20] as models trained using HICs data may not perform well in these cohorts due to biases in data acquisition and population demographics. For instance, suppose the United States, Canada, and Brazil collaborate to train an AI model using their combined data. Upon deployment, however, the model performs poorly specifically for the Brazilian population, despite Brazil's contribution of data. In this scenario, the model's overall good performance does not generate equal benefits for all three countries.

Federated learning (FL) [15], a popular distributed learning technique, promotes global collaboration for AI model development by training models in parallel without sharing data with a central server. Thereby, it facilitates access to broader and more diverse datasets encompassing various patient demographics, addressing data-sharing concerns and creating a model that is expected to perform well for every site participating in the training process [5]. However, despite these benefits, aggregating models trained in parallel poses considerable challenges related to two key factors: disparities in local dataset sizes and data heterogeneity across sites. These challenges are particularly pronounced for institutions in remote, underserved, and LMICs, where limited data availability could restrict their participation and reduce the potential benefits of the final FL model for their population. This has been shown by Li et al. [14] who analyzed various heterogeneity scenarios such as class, feature, and quantity imbalances across four distinct FL implementations (*i.e.*, FedAvg, FedProx, FedNova, and SCAFFOLD) using imaging and tabulated datasets. Although these FL implementations aggregate the locally trained models differently, their results showed that none of these implementations consistently outperforms the others across the evaluated scenarios, with the most severe performance drop observed for the class imbalance scenario. Moreover, Chang et al. [7] found that data heterogeneity across sites results in a biased global FL model. This may be because biases from sites with imbalanced data (*i.e.*, class, feature, and quantity) are inadvertently embedded within a subset of model parameters, causing the global model's dependency on sensitive attributes to increase during training. Their analysis revealed that this problem primarily emerges in the aggregation step, putting FL at a disadvantage compared to standard centralized models in the context of model fairness.

Compared to FL, the distributed learning method known as the travelling model (TM) [18] offers similar general advantages. It effectively addresses privacy concerns and issues related to dataset size and diversity without relying on an aggregation process. In the TM setup, a single model undergoes sequential training across multiple sites. This allows sites, even with very limited data, to actively engage and contribute to distributed learning training, mitigating the risk of sites with small sample sizes overfitting their data [18]. The participation of sites with limited data is crucial in advancing healthcare equity globally, as these centers may serve populations with unique characteristics. For example, when institutions in LMICs participate in AI model training, we gain valuable insights into how specific diseases manifest in their populations, thus helping to bridge the gap in research that traditionally relies on data from HICs. A recent study [19] showcased the effectiveness of the TM for Parkinson's disease (PD) classification using data from 83 sites with notable heterogeneity in the local datasets. However, [19] primarily focuses on the model's overall performance, a typical paradigm in evaluating and optimizing AI-based medical image models, which may overlook and hide important effects for site-specific populations. Consequently, further investigation is required to determine if all sites contributing data for TM training can benefit equally from the final model. Such an analysis can help advance research towards developing more equitable AI-based medical image models, preventing biased learning and motivating widespread contributions to TM development.

Inspired by the TM approach for PD classification, using the same data [19] and recognizing differences in PD manifestation based on sex and age [16], in this work, we explore how these factors influence the model's performance for each of the 83 sites engaged in the TM training. Specifically, our work evaluates and quantifies site-specific performance using the TM approach. This is done by analyzing the number of misdiagnoses (*i.e.*, false positive rate (FPR) and false negative rate (FNR)) per site. Therefore, the aim of this work is to emphasize the importance and establish a framework for site-specific analysis to promote better health equity in distributed learning, addressing the following questions: **Q1:** At an aggregate level (*i.e.*, population fairness), are there trends in misdiagnosis among population characteristics (*e.g.*, sex, age)? **Q2:** Which sites have the highest misdiagnosis rates? Are trends consistent with those seen at an aggregate level, or do site-specific characteristics (*e.g.*, number of datasets, class prevalence) lead to unequal distribution of benefits?

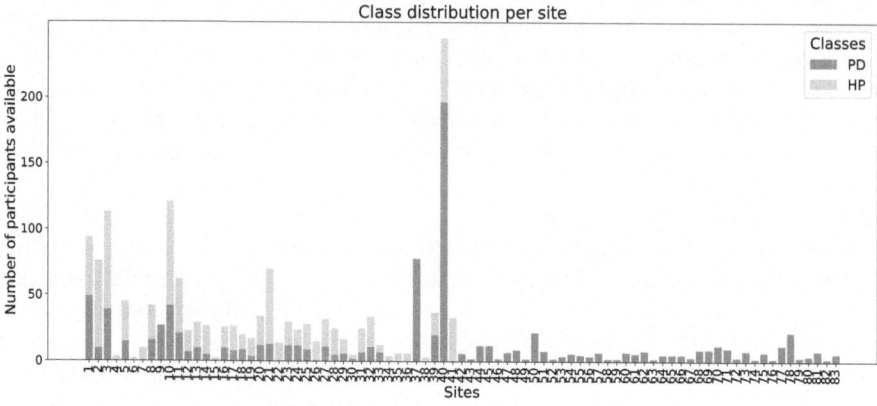

Fig. 1. Population characteristics per site. PD = Parkinson's disease, and HP = healthy participant.

2   Material and Methods

2.1   Data

We utilized the multi-site PD database comprised of 1817 T1-weighted magnetic resonance imaging (MRI) scans of the brain from 867 patients with PD and 950 healthy subjects obtained from 83 sites globally [1–4,6,8,10–13,22,23]. This database is unique for its diversity. Specifically, 54% of the sites provide only a limited number of datasets (*i.e.*, less than ten, see Fig. 1), and a considerable number of sites provide data with class considerable imbalance (see Fig. 1) as well as disparities in sex and age distributions (see supplementary material Fig 1 and 2). Such diversity reflects real-world scenarios, allowing investigation into whether all sites benefit from contributing data, which would have important implications for sites in underserved and remote regions.

The image preprocessing includes skull-stripping, resampling to an isotropic resolution of 1 mm, bias field correction, affine registration to an atlas [24], and cropping as described in [19].

2.2   Model Architecture

We utilized the same convolutional neural network as used in [19] consisting of seven blocks: The first five blocks included a 3D convolutional layer with $3 \times 3 \times 3$ kernel filters, batch normalization, $2 \times 2 \times 2$ max pooling, and ReLU activation.

Block six contained a 3D convolutional layer with $1 \times 1 \times 1$ kernel filters, along with batch normalization and ReLU activation. The seventh block comprised a 3D average pooling layer, a dropout layer with a 0.2 rate, and a flattening layer. Following this, a single dense layer with a sigmoid activation function was employed for the binary classification, distinguishing between patients with PD and healthy participants.

2.3   Training

We employed the TM approach for PD classification presented in [19] as the basis of this work. Briefly described, the TM method used in this work involved sequential training of a single model across multiple sites following a predetermined travel sequence dictating the order of visits. First, the model was trained using data available at the first site before moving to subsequent sites for further training using locally accessible data. This iterative process continued until training at all sites was completed, constituting one training cycle. Following this, a new travel sequence was randomly defined to introduce cycle-to-cycle variability, simulating the batch shuffling process commonly used in centralized approaches. Multiple cycles were conducted to improve the final model's performance. Using a grid search, it was determined that the optimal final model, as defined by the highest test accuracy across the 5-folds, needed 44 cycles (see supplementary material Fig 3). During training, a batch size of five was used when five or more datasets were available locally. The batch size was adjusted for sites with fewer than five datasets to match the available dataset size. The Adam optimizer started with an initial learning rate of 0.0001 and used an exponential decay rate with each cycle. Although the training was conducted on a single computer equipped with an NVIDIA GeForce RTX 3090 GPU, it strictly followed to the TM concept by retrieving data from only one site at a time for each epoch.

2.4   Evaluation

We quantitatively assessed the TM performance for each site by measuring the false positive rate (FPR) and false negative rate (FNR), which practically represent misdiagnosis. The FPR measures the proportion of healthy participants misclassified as patients with PD. Conversely, the FNR quantifies the proportion of patients with PD misclassified as healthy participants. To address **Q1**, we analyzed the misclassified datasets regardless of their sites (population fairness), focusing on sex and age. To investigate **Q2**, we first categorized FPR and FNR into four groups for each site: none (0%), low (up to 30%), medium (between 30–70%), and high (above 70%). Subsequently, we determined and grouped class representation categories as follows: HP only (site contributed only healthy participants' datasets), HP majority (site contributed more healthy participants than patients datasets), same (site contributed equal numbers of participants and patients datasets), PD majority (site contributed more patients than healthy participants datasets), and PD only (site contributed only patient datasets).

Next, we calculated the male ratio (fraction of males in a site's dataset) and the ratio of old participants (fraction of 65+ years old (mean age) in a site's dataset).

3   Results

The TM achieved an overall accuracy of 74%, similar to the results reported in [19]. However, we noted model performance was variable across sites, indicating that they did not all benefit equally.

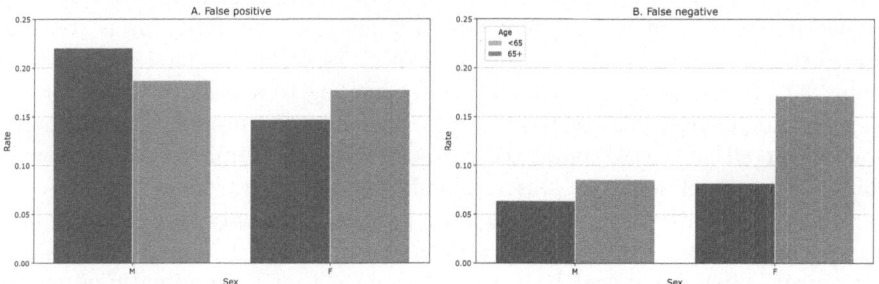

Fig. 2. Misdiagnosis percentages for specific sex-age groups. F = females and M = males.

3.1   Population Fairness Analysis

Table 1. Overall misdiagnosis considering sex and age as individual factors

	Males	Females	Older	Younger
FPR	20%	16%	19%	18%
FNR	7%	12%	7%	12%

At the aggregated level, the TM approach yielded a false positive rate (FPR) of 35% and a false negative rate (FNR) of 19%. Overall, Table 1 indicates that erroneous positive diagnoses are more prevalent among males (20%) and older individuals (19%), whereas females (12%) and younger individuals (12%) are more likely to receive erroneous negative diagnoses.

A deeper examination considering the combination of sex and age reveals a similar trend. Older males (22%) are more prone to false positive diagnoses, while younger females (17%) are more susceptible to underdiagnosis (see Fig. 2). This pattern may be attributed to the sex-dimorphic nature of PD, with a higher prevalence of the disease observed in males and older individuals [16].

3.2   Site-Based Analysis

In the population-level analysis, an established method for analyzing fairness in AI models, only sex and age were considered as factors that could influence misdiagnosis. Assuming that these factors were solely responsible for bias, we would expect that sites positioned in the center of Fig. 3 would show minimal levels of FPR and FNR, as these variables would be balanced across those sites. However, our findings reveal that many sites in this position still exhibit high levels of FPR. It is observed that sites with higher FPR and FNR (red shapes) both tend to have a greater proportion of males (x-axis), particularly concentrated between 60% and 80%. Additionally, the datasets at most of these sites predominantly include younger individuals (y-axis, where 0.5 corresponds to 65 years old). These observations deviate slightly from the population fairness analysis, indicating that additional site-level factors impact misdiagnosis rates.

Further analysis revealed that local dataset size and class representation also contribute to high levels of misdiagnosis. Although it might be expected that a model trained using distributed learning would struggle with smaller datasets, the presence of numerous sites represented by small green shapes suggests that dataset size alone does not influence misdiagnosis in this work. Instead, the imbalance in class representation emerges as the critical factor affecting elevated FPR and FNR, with most sites providing more PD cases (square shape) than healthy cases. Another indication that class imbalance is a key factor can be observed when analyzing the sites with none or low levels of FPR and FNR. Many of these sites have a single class representation in their dataset, providing only healthy participants (green circles) or only patients with PD (green diamonds).

4   Discussion

This work introduces a framework for site-based analysis in distributed learning setups, revealing that class representation within a site greatly influences the misdiagnosis rates of the travelling model (TM). Our findings revealed high rates of misdiagnosis at sites that contributed large datasets with a higher prevalence of Parkinson's disease (PD) cases and a higher proportion of male participants. This site-specific analysis provided us with deeper insights into how factors unique to each site affect the performance of the model across different data contributors. Identifying the potential interactions among these factors would not have been achievable through aggregate population fairness analysis alone nor through a simplistic evaluation based solely on the overall accuracy of the final model. Consequently, this work underscores the importance of developing a comprehensive understanding of model behavior, particularly in distributed learning setups, to ensure that all data contributors are able to benefit from the final model. Given these results, we encourage developers to evaluate and optimize models beyond mere overall accuracy, a crucial effort for advancing the development of AI-based medical image models that can effectively address global healthcare disparities.

This work provides valuable insights how to improve AI model development in the distributed learning domain. In the context of the travelling model, the approach specifically investigated in this work, understanding the variation in benefits across different sites and populations may assist developers in exploring enhanced strategies for defining the travel sequence, as the order in which the model utilizes data during training has been shown to impact deep learning models performance in prior work [9]. Likewise, considering the distinct context of each contributing site and its population could also aid developers in refining aggregation procedures in the federated learning approach. In both scenarios, the strategy should aim to explore local heterogeneity to achieve a more equitable distribution of benefits among all sites.

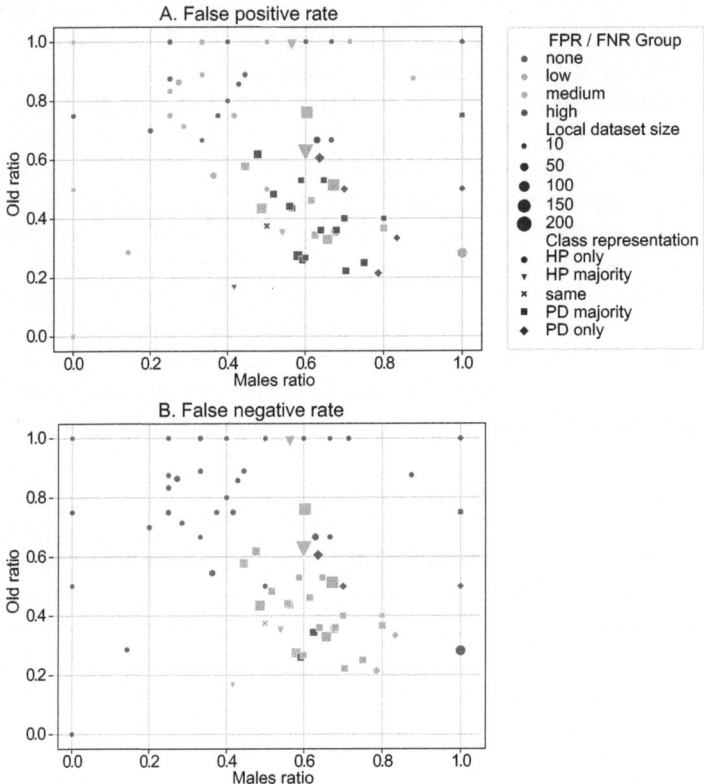

Fig. 3. The plots illustrate the false positive rate (A) and false negative rate (B) alongside data availability per site (size of shapes), the male ratio (x-axis: 0.0 for only females, 1.0 for only males), the age distribution (y-axis: below 0.5 for younger than 65 years, above 0.5 for older), and the class representation (shapes) for each model based on data from each site.

The insights from our work extend beyond the domain of distributed learning, offering benefits to various AI domains that utilize multi-site datasets that span various regions with differences in imaging technology, patient cohorts, and medical guidelines and procedures. Assuming every site contributing data will receive the final model, it is crucial to ensure that no site is disadvantaged by poor model performance if the model has not learned to generalize to its data. If this is the case, our site-specific analysis, which identifies which subpopulations are often misdiagnosed, could be used to gain a deeper insight into what biases or shortcuts are being potentially used by the model [21]. For example, multi-site databases often encounter non-biological discrepancies, such as image acquisition protocols and scanner variations. These discrepancies can be particularly notable when combining data from high-income countries (HICs) and low- and middle-income countries (LMICs) to train AI models. While HICs are more likely to have access to newer scanners, LMICs often depend on older machines [17]. In this context, our analysis could be extended to identify how scanner types impact the model performance at individual sites. This analysis could facilitate the development of targeted bias mitigation or harmonization techniques to enhance overall and site-specific generalizability.

5 Conclusion

This work introduces, for the first time, an analysis of the disparities in benefits among sites employing the TM approach. While such disparities are evident, this framework analysis provides novel insights that are crucial for fostering more equitable model development. This is essential for mitigating the risks of misdiagnosis associated with the use of AI-driven healthcare solutions, particularly for patient populations in remote or under-resourced settings.

References

1. Open science, to accelerate discovery and deliver cures — the neuro - mcgill university. https://www.mcgill.ca/neuro/open-science
2. Openneuro. https://openneuro.org/datasets/ds000245/versions/00001
3. Parkinson's progression markers initiative. https://www.ppmi-info.org/
4. Acharya, H.J., Bouchard, T.P., Emery, D.J., Camicioli, R.M.: Axial signs and magnetic resonance imaging correlates in Parkinson's disease. Can. J. Neurol. Sci./Journal Canadien des Sciences Neurologiques **34**, 56–61 (2007). https://doi.org/10.1017/S0317167100005795
5. Annas, G.J.: HIPAA regulations - a new era of medical-record privacy? N. Engl. J. Med. **348**, 1486–1490 (2009). https://doi.org/10.1056/NEJMLIM035027
6. Badea, L., Onu, M., Wu, T., Roceanu, A., Bajenaru, O.: Exploring the reproducibility of functional connectivity alterations in Parkinson's disease. PLoS ONE **12**, e0188196 (2017). https://doi.org/10.1371/journal.pone.0188196
7. Chang, H., Shokri, R.: Bias propagation in federated learning. In: International Conference on Learning Representations (2023). http://arxiv.org/abs/2309.02160
8. Duchesne, S., et al.: The Canadian dementia imaging protocol: harmonizing national cohorts. J. Magn. Reson. Imaging **49**, 456–465 (2019). https://doi.org/10.1002/jmri.26197

9. Ganesh, P., Chang, H., Strobel, M., Shokri, R.: On the impact of machine learning randomness on group fairness. In: ACM International Conference Proceeding Series, vol. 12, pp. 1789–1800 (2023). https://doi.org/10.1145/3593013.3594116

10. Hanganu, A., et al.: Mild cognitive impairment is linked with faster rate of cortical thinning in patients with Parkinson's disease longitudinally. Brain **137**, 1120–1129 (2014). https://doi.org/10.1093/brain/awu036

11. Jack, C.R., et al.: The Alzheimer's disease neuroimaging initiative (ADNI): MRI methods. J. Magn. Reson. Imaging **27**, 685–691 (2008). https://doi.org/10.1002/jmri.21049

12. LaMontagne, P.J., et al.: Oasis-3: longitudinal neuroimaging, clinical, and cognitive dataset for normal aging and Alzheimer disease. medRxiv p. 2019.12.13.19014902 (2019). https://doi.org/10.1101/2019.12.13.19014902

13. Lang, S., et al.: Network basis of the dysexecutive and posterior cortical cognitive profiles in Parkinson's disease. Mov. Disord. **34**, 893–902 (2019). https://doi.org/10.1002/mds.27674

14. Li, Q., Diao, Y., Chen, Q., He, B.: Federated learning on non-IID data silos: an experimental study (2021). https://arxiv.org/abs/2102.02079v4

15. McMahan, H.B., Moore, E., Ramage, D., Hampson, S., y Arcas, B.A.: Communication-efficient learning of deep networks from decentralized data (2016)

16. Meoni, S., Macerollo, A., Moro, E.: Sex differences in movement disorders. Nat. Rev. Neurol. **16**(2), 84–96 (2020). https://doi.org/10.1038/s41582-019-0294-x

17. Mollura, D.J., et al.: Artificial intelligence in low- and middle-income countries: innovating global health radiology. Radiology **297**(3), 513–520 (2020)

18. Souza, R., Mouches, P., Wilms, M., Tuladhar, A., Langner, S., Forkert, N.D.: An analysis of the effects of limited training data in distributed learning scenarios for brain age prediction. J. Am. Med. Inform. Assoc. **30**, 112–119 (2022). https://doi.org/10.1093/jamia/ocac204

19. Souza, R., et al.: A multi-center distributed learning approach for Parkinson's disease classification using the traveling model paradigm. Front. Artif. Intell. **7** (2024). https://doi.org/10.3389/frai.2024.1301997

20. Souza, R., Stanley, E.A., Forkert, N.D.: On the relationship between open science in artificial intelligence for medical imaging and global health equity. LNCS, vol. 14242, pp. 289–300 (2023). https://doi.org/10.1007/978-3-031-45249-9_28/FIGURES/1

21. Souza, R., et al.: Image-encoded biological and non-biological variables may be used as shortcuts in deep learning models trained on multisite neuroimaging data. J. Am. Med. Inform. Assoc. (2023). https://doi.org/10.1093/jamia/ocad171

22. Sudlow, C., et al.: UK biobank: an open access resource for identifying the causes of a wide range of complex diseases of middle and old age. PLoS Med. **12**, e1001779 (2015). https://doi.org/10.1371/journal.pmed.1001779

23. Talai, A.S., Sedlacik, J., Boelmans, K., Forkert, N.D.: Utility of multi-modal MRI for differentiating of Parkinson's disease and progressive supranuclear palsy using machine learning. Front. Neurol. **12**, 648548 (2021). https://doi.org/10.3389/fneur.2021.648548

24. Xiao, Y., et al.: A dataset of multi-contrast population-averaged brain MRI atlases of a Parkinson's disease cohort. Data Brief **12**, 370–379 (2017). https://doi.org/10.1016/j.dib.2017.04.013.   https://linkinghub.elsevier.com/retrieve/pii/S2352340917301452

Cycle-GANs Generated Difference Maps to Interpret Race Prediction from Medical Images

Lakshika Rathi[1], Giacomo Nebbia[2]([✉]), Ken Chang[3], Sourav Kumar[2],
Aarushi Gupta[1], Syed Rakin Ahmed[5], Jay Patel[6], Christopher Clark[2],
Yoga Advaith Veturi[2], Aaron Coyner[7], Aakanksha Rana[6],
Christopher Bridge[8], Stephen McNamara[2], J. Peter Campbell[7],
Matthew Li[4], Jayashree Kalpathy-Cramer[2], and Praveer Singh[2]

[1] Indian Institute of Technology, Delhi, Delhi, India
[2] University of Colorado Anschutz Medical Campus, Aurora, CO, USA
giacomo.nebbia@cuanschutz.edu
[3] Athinoula A. Martinos Center for Biomedical Imaging, Boston, MA, USA
[4] Harvard University, Cambridge, MA, USA
[5] Massachusetts Institute of Technology, Cambridge, MA, USA
[6] Oregon Health And Science University, Portland, OR, USA
[7] Massachusetts General Hospital, Boston, MA, USA
[8] University of Alberta, Edmonton, AB, Canada

Abstract. Recent research has revealed the remarkable ability of artificial intelligence (AI) to identify features related to an individual's self-reported race in medical images, but what such features may be remains an unanswered question. In this work, we aim to identify image regions relevant to race prediction. We argue that previous methods toward this goal (namely, occlusion maps) are not sufficient as they are unable to locate such regions, and we propose to use Cycle-GANs as an alternative. Specifically, we train a Race-specific Cycle-GAN to artificially transform images from patients of one race to images from patients of a different race. We then obtain difference maps by computing the pixelwise difference between original and transformed image. Difference maps highlight pixels whose values are crucial for an image to be considered as belonging to a patient of a specific race. Additionally, we examine whether such regions are gender dependent by subgrouping our analysis for male and female patients. We show how difference maps are able to identify relevant image regions when previously introduced methods fail, and that, while some differences do exist between genders, the relevant regions mainly overlap.

Keywords: race prediction · interpretability · difference maps

E. Puyol-Antón et al. (Eds.): FAIMI 2024/EPIMI 2024, LNCS 15198, pp. 129–139, 2025.
https://doi.org/10.1007/978-3-031-72787-0_13

1   Introduction

Among radiologists, there exists a tacit belief that no signs of a patient's race can be found in medical imaging [7]. However, recent studies have demonstrated that AI models can predict race with impressive accuracy from several types of imaging [3,7], including radiographs [22], mammograms [3], and fundus photographs [5]. While previous work has investigated initial hypotheses related to how such models are able to predict race [7], no definite answer has been proposed. In fact, previous results show that an algorithm's performance remains virtually unchanged when utilizing lower-resolution images, when filtering out high frequencies, and when any part of the image is masked out [7]. This masking approach aims to identify image regions a classification model relies on to make a prediction by masking out (i.e., occlude) different parts of an image. By recording how the probability of prediction changes when each image region is masked, we obtain an occlusion map highlighting areas a model significantly relies on to make the correct class prediction. Our work is related to occlusion maps as it also aims to compute image maps highlighting relevant regions. We will show how occlusion maps fail at this task, while our proposed difference maps succeed.

In this work, we address this lack of methods capable of identifying relevant image features by introducing a new way to identify image regions relevant to the race prediction task. Specifically, we compute pixel-level difference maps between an image and an artificially altered version obtained from a Cycle-GAN [26]. In detail, we propose a Race-specific Cycle-GAN (RC-GAN), trained to alter an image from a patient of one race so that it would be more similar to images from patients of a different race. Given a Convolutional Neural Network (CNN) capable of predicting race from the original images, the difference map for a given image represents pixels whose values are highly relevant to race prediction if the CNN's race predictions for the original and the altered images are different. In other words, changing the values of those pixels "tricks" the race prediction CNN into predicting a different race than the one predicted for the original image.

Our experiments on radiology data show the superiority of our proposed methods to the previously introduced occlusion maps by being able to identify highly relevant image regions that occlusion maps are unable to uncover. In addition, we investigate whether such relevant regions are gender-dependent by comparing difference maps for male patients with those for female patients. While some differences do exist, the relevant regions are very similar.

The remainder of the paper is organized as follows: Sect. 2 introduces related work, Sect. 3 explains how to obtain difference maps from our Race-Specific Cycle-GAN, Sect. 4 details our experimental design, Sect. 5 presents our results, and Sect. 6 draws our conclusions.

2   Related Work

2.1   Race Prediction

Deep learning methods have shown the ability to predict self-reported race from a variety of medical images [3,7], even if specialists did not believe race-related features existed in such images [7]. This finding is significant because previous research has also shown how automated diagnosis/screening tools can be biased based on a patient's race [15,20]. This is problematic, and solutions have been put forth to try to ameliorate this problem and train fairer models [4,22,24].

In addition to the risk of introducing bias in deep learning models, their ability of predicting a patient's self-reported race poses a threat to that patient's privacy, exposing them to the risk of re-identification [12,16]. For instance, in ophthalmology, models have been shown to be able to predict not only self-reported race [5], but also gender [17], age [10], and a number of different systemic variables [2]. Compounding all these pieces of information may lead to unwarranted re-identification of patients; consequently, it is paramount to understand how these models are able to predict such demographic information and how to possibly counter such ability. In this work, we contribute to the former by visualizing image regions relevant to the task of predicting self-reported race.

2.2   Interpretability Methods

Due to the fact that deep learning methods learn features as well as a decision boundary to perform classification, researchers have been interested in understanding what such features are. Focusing on computer vision, previously introduced methods visualize such features either directly [6,14] or by generating maps overlaid over the input images that would highlight image regions relevant to the classification task [19,23,23,25]. These methods differ based on how the maps are generated: occlusion maps sequentially mask out image patches and register the change in probability for a given class of interest [23], while saliency maps [21] and gradient activation maps [19,25] use information about gradients to highlight image regions that are highly influential for the prediction task. Both occlusion maps [7] and gradient-based saliency maps [1] have been shown to be inadequate for localization of areas of interest in medical imaging. Alternatively, Generative Adversarial Networks (GANs) have also been proposed as an interpretability tool by training them to generate synthetic images from real ones that could "trick" a model into changing its prediction when presented with the synthetic image [11,13,18]. Similar to previous approaches, we also use a GAN model to generate synthetic images, but, while previous methods stop at image generation and rely on visual inspection of the generated images to discern relevant features and regions, we use the synthetic images to generate maps of regions of the input images that are relevant to our classification task.

3   Methods

3.1   Occlusion Maps

To compute an occlusion map [23], an image is divided into a grid of patches. One patch at a time is masked out by setting all pixels to the same value, and the partially masked image is passed to the classification model. The prediction probability is recorded and assigned to that patch and then subtracted from the probability predicted when the whole, unmasked image is passed through the model. Once this process is completed for all patches, we obtain a heatmap showing by how much each patch causes the predicted probability to vary. Values of high magnitude represent image regions highly relevant for the task, while values of low magnitude represent rather unimportant image regions. Figure 1 depicts how to compute an occlusion map for a 4x4 grid.

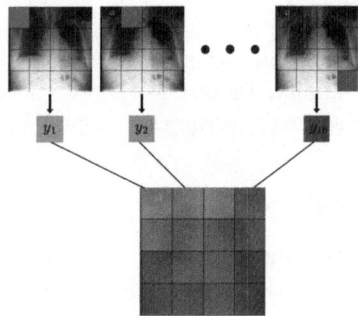

Fig. 1. Example of how to compute an occlusion map for a chest x-ray image with a 4x4 grid. Each patch of the grid is, in turn, masked out, and the masked images is passed through the network. The difference between the probabilities predicted from the masked image and from the original image is recorded for each patch. High-magnitude values (in purple) represent image regions highly relevant to the predicted task, while low-magnitude values (in yellow) represent regions that are rather unimportant. (Color figure online)

3.2   Difference Maps from Cycle-GAN

The Cycle-GAN model [26] was introduced for the task of image translation: translate an image from domain X to one from domain Y (in our study, X and Y correspond to two different self-reported races). While previous image translation models require the presence of paired training samples (i.e., two versions of the same image: one belonging to domain X and one to domain Y), Cycle-GAN relaxes this constraint by introducing a cycle consistency loss. Such loss enforces the fact that if image $x \in X$ is translated to image $y \in Y$ through the generator G (i.e., $y = G(x)$) and image $y' \in Y$ is translated to image $x' \in X$ through generator F, then $F(G(x)) \approx x$ and $G(F(y')) \approx y'$. This idea is well suited for translating images between self-reported races since it is impossible to have paired data in this scenario (e.g., an image belonging to both a patient with self-reported race white and self-reported race black).

Figure 2 depicts our proposed Race-specific Cycle-GAN framework and how we compute difference maps. After training our Cycle-GAN model, we are able to convert images from one group to images of the other group (in both directions). In our case, we will alter images from patients with self-reported race white into images for patients with self-reported race black (and vice-versa). For each original-altered image pair, we can compute the pixel-wise difference and obtain a difference map where high-magnitude values represent pixels that had to be changed a lot to transform the image. Such high-value pixels thus represent the image regions highly relevant to the race prediction task, which we are interested in identifying.

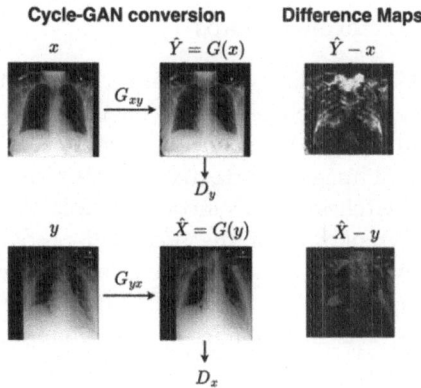

Fig. 2. Left: structure of our Race-specific Cycle-GAN. An image from a patient of race x is altered by generator G_{xy} into an altered image looking like those from patients of race y. Similarly, an image from a patient of race y is transformed by a generator G_{yx}. Two discriminator networks D_x and D_y are trained to distinguish whether the altered image for race x (or y) belongs to the original race. Right: difference maps obtained by pixel-wise difference between the altered image and the original image. High values represent pixels that need to be altered a lot to "change" an image from one race to the other.

4   Experiments

4.1   Dataset

We use the publicly available MIMIC-CXR [9] dataset including chest x-rays and clinical data comprising 227,835 radiographic studies performed at the Beth Deaconess Medical Center in Boston, MA. Clinical data includes information on self-reported race; specifically, we extracted 150,370 images for self-reported white patients and 36,574 images for self-reported black patients. We split this dataset into training/validation/test with a 6:1:3 ratio (at patient-level) for model training and evaluation.

4.2   Race Prediction

We train a CNN model to predict race from chest x-rays and compute occlusion maps for patients whose race is correctly classified and difference maps only for patients whose race prediction changes to the incorrect race when the altered image is used as input. We exclude images whose race prediction does not change when the altered image is used as input because, if the model keeps predicting the race of the original image, it means that the changes made by the RC-GAN are insufficient to cause a reversal of the prediction, and the altered image regions may thus not be relevant to race prediction.

4.3   Occlusion and Difference Maps

After training an RC-GAN model, we generate original-altered image pairs for all the available images. For each pair, each image is passed through the race prediction network, and the predicted race is recorded. We consider only pairs where race for the original image is correctly predicted and where the prediction on the altered image switches to the incorrect race. We show occlusion and difference maps for different bins of the predicted probability for the altered image. Such probability indicates the confidence with which the model flips its prediction after the original image is altered. Finally, we investigate whether the difference maps differ based on gender by sub-grouping difference maps and visually comparing them.

4.4   Implementation

We implemented our race classification CNN using code from this GitHub repository (https://github.com/Emory-HITI/AI-Vengers/tree/main). The race classification network architecture is a ResNet34 [8] with an added fully connected layer at the end to perform binary classification. We trained the model using the categorical cross-entropy loss and early stopping if the validation loss does not decrease for 4 consecutive epochs. All hyperparameters were kept unchanged from the original code (e.g., learning rate 0.001, decreased by a factor of 10 if the validation loss does not decrease for more than 2 epochs, with a minimum learning rate of 10^{-5}). To train our RC-GAN, we used the official CycleGAN code base (https://github.com/junyanz/pytorch-CycleGAN-and-pix2pix). We trained for 100 epochs, keeping the generator frozen for the first 5 epochs, and saving the model that reached the highest accuracy of conversion from one race to the other (in our case, at epoch 35). All hyperparameters were kept unchanged (e.g., learning rate 0.0002).

5 Results

5.1 Occlusion Maps

Figure 3 shows the occlusion maps for the original images when the predicted race is correctly identified as self-reported white (left) and self-reported black (right) and the incorrect race from the RC-GAN altered image is predicted with probability > 0.99 (total of 4,943 and 3,069 images for self-reported white and black, respectively). This is done for fair comparison between occlusion and difference maps. The shown maps are the pixel-wise average across the individual occlusion maps for each image. We notice how the probabilities change by up to 0.1, showing the inability of this method to highlight relevant areas.

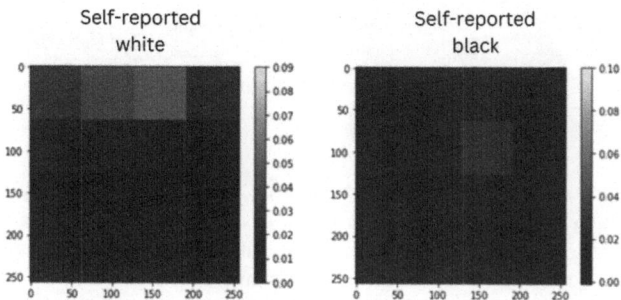

Fig. 3. Occlusion maps (averaged across patients) for original images where the predicted probability of the correct class is above 0.99. Left: self-reported white race. Right: self-reported black race.

5.2 Difference Maps

Figure 4 shows the difference maps for different bins of predicted probability for the incorrect class on the altered images (p_{black} and p_{white} for self-reported white and black patients, respectively). The shown difference maps are the pixel-wise average across the individual difference maps for each image. The number of images (for self-reported white and black race, respectively) for each probability bins are: $p > 0.99$ 4,943 and 3,069, $p \in [0.9, 0.99]$ 976 and 348, $p \in [0.5, 0.9)$ 709 and 343, and $p < 0.5$ 1,643 and 1,670. We notice how the maps for different probability bins appear to identify the same regions, but the differences in pixel values in those regions decrease, supporting the claim that those regions are highly relevant to race prediction (i.e., the more we change those pixel values, the more the race prediction model will be swayed). Such areas are also localized in the neck region and in the lungs and arms. Interestingly, pixels in the neck region (and the bottom corners of the image) and the lungs and arms are changed in opposite directions (as depicted by the areas being of two colors). As expected, the maps for race conversion are opposite in the two directions; lowering pixels highly relevant for self-reported white prediction (blue areas in the bottom row) increases the probability of the model flipping its prediction if a patient is white (red areas in the top row).

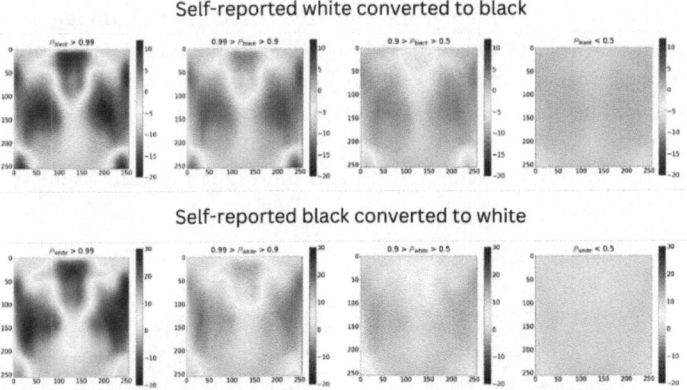

Fig. 4. Difference maps (averaged across patients) for different bins of probability predicted for the incorrect class on the altered images. Top: difference maps for images for patients with self-reported race white. Bottom: difference maps for images for patients with self-reported race black. The right-most column represents maps for images whose alterations where not sufficient to cause the race prediction model to change its prediction.

Finally, we sub-group the highest probability bin (i.e., $p > 0.99$) based on gender, obtaining the maps shown in Fig. 5. In general, the neck and lung

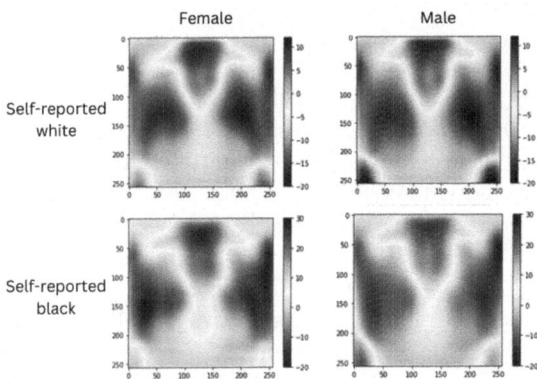

Fig. 5. Difference maps (averaged across patients) for female and male gender for the patients with probability of incorrect race on the altered image > 0.99. Top: patients whose race is incorrectly classified as self-reported white. Bottom: patients whose race is incorrectly classified as self-reported black.

regions remain relevant between genders, although the map intensity does change between genders, especially for patients with self-reported white race (top row).

6 Conclusions

In this work, we aimed to address the lack of effective interpretability methods to identify image regions relevant for a prediction task. Specifically, we focused on race prediction from chest x-rays. We argued that previously introduced region identification methods (namely, occlusion maps) are unable to identify such regions, and we conducted experiments to show this is true in our datasets. We thus introduced a novel identification methods: difference maps from a Race-specific Cycle-GAN (RC-GAN), which proved successful at identifying relevant image regions where occlusion maps fail. Finally, we identified minor differences in relevant image regions for race predictions based on a patient's gender.

While our results are compelling, our study has some limitations. First, difference maps require two models to be trained: a race prediction network and a Cycle-GAN (occlusion maps only require the former); the interpretability gains thus seem to come at the cost of increased computational requirements. Second, we only tested our method on one prediction task; a more comprehensive evaluation on a plethora of tasks of interest will provide more insight in the value of our approach. Third, we ran experiments on one public dataset; the inclusion of additional datasets is a potential direction for future work. Finally, we only tested our difference maps against occlusion maps; additional methods exist in explainable AI (like the gradient-based methods we mentioned [19,25]) that could be compared against our difference maps.

References

1. Arun, N., et al.: Assessing the trustworthiness of saliency maps for localizing abnormalities in medical imaging. Radiol. Artif. Intell. **3**(6), e200267 (2021)
2. Babenko, B., et al.: A deep learning model for novel systemic biomarkers in photographs of the external eye: a retrospective study. Lancet Digit. Health **5**(5), e257–e264 (2023)
3. Banerjee, I., et al.: Reading race: AI recognises patient's racial identity in medical images. arXiv preprint arXiv:2107.10356 (2021)
4. Burlina, P., Joshi, N., Paul, W., Pacheco, K.D., Bressler, N.M.: Addressing artificial intelligence bias in retinal diagnostics. Transl. Vis. Sci. Technol. **10**(2), 13–13 (2021)
5. Coyner, A.S., et al.: Association of biomarker-based artificial intelligence with risk of racial bias in retinal images. JAMA Ophthalmol. **141**(6), 543–552 (2023)
6. Erhan, D., Bengio, Y., Courville, A., Vincent, P.: Visualizing higher-layer features of a deep network. Univ. Montreal **1341**(3), 1 (2009)
7. Gichoya, J.W., et al.: AI recognition of patient race in medical imaging: a modelling study. Lancet Digit. Health **4**(6), e406–e414 (2022)

8. He, K., Zhang, X., Ren, S., Sun, J.: Deep residual learning for image recognition. In: Proceedings of the IEEE Conference on Computer Vision and Pattern Recognition, pp. 770–778 (2016)

9. Johnson, A.E., et al.: MIMIC-CXR, a de-identified publicly available database of chest radiographs with free-text reports. Sci. Data **6**(1), 317 (2019)

10. Khan, N.C., et al.: Predicting systemic health features from retinal fundus images using transfer-learning-based artificial intelligence models. Diagnostics **12**(7), 1714 (2022)

11. Lang, O., et al.: Explaining in style: training a GAN to explain a classifier in stylespace. In: Proceedings of the IEEE/CVF International Conference on Computer Vision, pp. 693–702 (2021)

12. Macpherson, M.S., Hutchinson, C.E., Horst, C., Goh, V., Montana, G.: Patient reidentification from chest radiographs: an interpretable deep metric learning approach and its applications. Radiol. Artif. Intell. **5**(6), e230019 (2023)

13. Narayanaswamy, A., et al.: Scientific discovery by generating counterfactuals using image translation. In: Martel, A.L., et al. (eds.) MICCAI 2020. LNCS, vol. 12261, pp. 273–283. Springer, Cham (2020). https://doi.org/10.1007/978-3-030-59710-8_27

14. Nguyen, A., Dosovitskiy, A., Yosinski, J., Brox, T., Clune, J.: Synthesizing the preferred inputs for neurons in neural networks via deep generator networks. In: Advances in Neural Information Processing Systems, vol. 29 (2016)

15. Parikh, R.B., Teeple, S., Navathe, A.S.: Addressing bias in artificial intelligence in health care. JAMA **322**(24), 2377–2378 (2019)

16. Raghu, V.K., Lu, M.T.: Chest radiographs: a new form of identification? Radiol. Artif. Intell. **5**(6), e230397 (2023)

17. Rim, T.H., et al.: Prediction of systemic biomarkers from retinal photographs: development and validation of deep-learning algorithms. Lancet Digit. Health **2**(10), e526–e536 (2020)

18. Schutte, K., Moindrot, O., Hérent, P., Schiratti, J.B., Jégou, S.: Using stylegan for visual interpretability of deep learning models on medical images. arXiv preprint arXiv:2101.07563 (2021)

19. Selvaraju, R.R., Cogswell, M., Das, A., Vedantam, R., Parikh, D., Batra, D.: Gradcam: visual explanations from deep networks via gradient-based localization. In: Proceedings of the IEEE International Conference on Computer Vision (ICCV) (2017)

20. Seyyed-Kalantari, L., Zhang, H., McDermott, M.B., Chen, I.Y., Ghassemi, M.: Underdiagnosis bias of artificial intelligence algorithms applied to chest radiographs in under-served patient populations. Nat. Med. **27**(12), 2176–2182 (2021)

21. Simonyan, K., Vedaldi, A., Zisserman, A.: Deep inside convolutional networks: visualising image classification models and saliency maps. In: Proceedings of the International Conference on Learning Representations (ICLR). ICLR (2014)

22. Wang, R., Kuo, P.C., Chen, L.C., Seastedt, K.P., Gichoya, J.W., Celi, L.A.: Drop the shortcuts: image augmentation improves fairness and decreases AI detection of race and other demographics from medical images. EBioMedicine **102** (2024)

23. Zeiler, M.D., Fergus, R.: Visualizing and understanding convolutional networks. In: Fleet, D., Pajdla, T., Schiele, B., Tuytelaars, T. (eds.) ECCV 2014. LNCS, vol. 8689, pp. 818–833. Springer, Cham (2014). https://doi.org/10.1007/978-3-319-10590-1_53

24. Zhang, B.H., Lemoine, B., Mitchell, M.: Mitigating unwanted biases with adversarial learning. In: Proceedings of the 2018 AAAI/ACM Conference on AI, Ethics, and Society, pp. 335–340 (2018)

25. Zhou, B., Khosla, A., Lapedriza, A., Oliva, A., Torralba, A.: Learning deep features for discriminative localization. In: Proceedings of the IEEE Conference on Computer Vision and Pattern Recognition, pp. 2921–2929 (2016)
26. Zhu, J.Y., Park, T., Isola, P., Efros, A.A.: Unpaired image-to-image translation using cycle-consistent adversarial networks. In: Proceedings of the IEEE International Conference on Computer Vision, pp. 2223–2232 (2017)

On Biases in a UK Biobank-Based Retinal Image Classification Model

Anissa Alloula[1]([✉])[iD], Rima Mustafa[1][iD], Daniel R. McGowan[2,3][iD], and Bartłomiej W. Papież[1][iD]

[1] Big Data Institute, University of Oxford, Oxford, UK
anissa.alloula@dtc.ox.ac.uk
[2] Department of Oncology, University of Oxford, Oxford, UK
[3] Department of Medical Physics and Clinical Engineering, Oxford University Hospitals NHS FT, Oxford, UK

Abstract. Recent work has uncovered alarming disparities in the performance of machine learning models in healthcare. In this study, we explore whether such disparities are present in the UK Biobank fundus retinal images by training and evaluating a disease classification model on these images. We assess possible disparities across various population groups and find substantial differences despite strong overall performance of the model. In particular, we discover unfair performance for certain assessment centres, which is surprising given the rigorous data standardisation protocol. We compare how these differences emerge and apply a range of existing bias mitigation methods to each one. A key insight is that each disparity has unique properties and responds differently to the mitigation methods. We also find that these methods are largely unable to enhance fairness, highlighting the need for better bias mitigation methods tailored to the specific type of bias.

Keywords: Machine Learning · Bias · UK Biobank · Retinal Imaging

1 Introduction and Related Work

Biases and Disparities in Machine Learning. An emerging concern in machine learning (ML) research is that strong overall performance may obscure critical disparities, leading to substantially inferior outcomes for certain subgroups. Examples of this unequal performance have been identified in clinical ML models, across a range of tasks and modalities such as skin lesion classification [3], brain Magnetic Resonance Imaging (MRI) reconstruction [10], cardiac MRI segmentation [19], and affecting various subgroups, from certain ethnic groups [16,19], to disadvantaged socioeconomic groups [24]. Not only do these biases harm the minority groups who are subject to them, but they also hinder

Supplementary Information The online version contains supplementary material available at https://doi.org/10.1007/978-3-031-72787-0_14.

the generalisability of the models to unseen population samples [21], constituting a major barrier to the implementation of ML models in clinical settings.

Existing Approaches to Address Such Biases. A line of research focused on preventing such disparities has consequently emerged. Bias mitigation can be conducted at various stages in the ML pipeline: during data collection, in the pre-processing stage, while the model is training, and/or in post-processing. Objectives vary between methods and can include boosting minimum performance [9], reducing gaps in performance (equalised odds [10]), or equalising the number of positive predictions across groups (demographic parity [6]). However, recent work has highlighted that despite the multitude of existing methods, the problem is far from solved. A benchmark from 2023, MEDFAIR, showed that across a range of medical tasks, no method consistently and significantly outperformed empirical risk minimisation (where there is no fairness objective) [33].

Problem Setting. In this study, we focus on the appearance of biases and their mitigation in retinal imaging-based models. Bias mitigation research has been lacking in this field, with, to the best of our knowledge, only two examples: work by Burlina et al. [5] and work by Coyner et al. [8], who tried to mitigate race-related disparities with synthetic data and data pre-processing, respectively. We build on this work by conducting the largest and most comprehensive exploration of disparities and mitigation methods in retinal imaging to date. We use retinal images from the UK Biobank (UKBB), an unparalleled medical database of over half a million UK adults [25]. We complement recent work which has identified selection bias in the UKBB [4,17,23,26] by considering other possible bias types and how they manifest in ML models. In addition to providing insights on understudied possible biases in retinal imaging, the use of this database allows us to consider what disparities remain when standardisation has been conducted, as the UKBB has undergone rigorous data acquisition and quality control protocols [2], such that all images were taken with the same type of OCT scanner [1]. Also, the breadth of data available in the UKBB allows us to specifically characterise different biases (including some which are rarely investigated).

Contributions. We train a retinal image hypertension classification model on images from more than 75,000 individuals and use this as a proxy task to understand possible biases. We find that our model has uneven performance across subgroups, including between images from different assessment centres. We explore possible reasons for these disparities among common factors such as data imbalance, image quality, unequal generalisation, and separations in the model's representations of different subgroups, and find that these do not necessarily hold true depending on the disparity. Finally, we find that no bias mitigation method manages to consistently improve the fairness of our model. This highlights the non-universality of existing bias mitigation methods and underscores the need for a framework to specifically characterise disparities and their causes, as well as to determine if and how to best minimise them.

2   Methods and Experimental Setup

Dataset and Pre-processing. We use 80,966 fundus retinal images from the right eye of 78,346 individuals in the UKBB. We exclude 1,874 images corresponding to participants who had subsequently withdrawn, who had "other", "preferred not to say", or "unknown" ethnicity, and those from one assessment centre which had fewer than 0.2% of images. The UKBB is particularly rich in available metadata, including age, body mass index (BMI), self-reported alcohol consumption, self-reported ethnicity, genetic ethnicity (gen_ethnicity), genetic sex, deprivation, medication, etc. We create categorical groupings for age (40–50, 50–60, 60–70, 70+), BMI (0–3 based on quartile), deprivation index (0–3 based on quartile), and self-reported ethnicity (White, mixed background, Asian background, or Black African background) to facilitate downstream analyses. We anonymise the names of the centres.

We also adjust diastolic and systolic blood pressure (BP) by +10 and +15 mm Hg, respectively, if individuals are taking hypertensive medication [28]. We classify individuals as having high blood pressure (hypertension) if: diastolic $BP \geq 80$ or systolic $BP \geq 130$ or if they are taking anti-hypertensive medication (according to the current guidelines [30]). This is the binary target variable our model aims to predict. Figure A1 shows some of the dataset characteristics.

Model Architecture and Training. We split data into train, validation, and test sets (0.8, 0.1, 0.1) stratifying by individuals. As in [18], we train an InceptionV3 Network [27] to classify a retinal image as belonging to a hypertensive or non-hypertensive individual. Table 1 shows specific implementation details.

Table 1. Implementation details.

Training strategy	Implementation
Network backbone	InceptionV3
Pre-training	ImageNet
Batch size	512
Image size	$3 \times 299 \times 299$
Augmentation	Random flip, rotation, crop, color jitter, Gaussian blur
Optimiser	Adam
Loss	Binary cross-entropy
Learning rate	0.0005
Learning scheduler	StepLR (gamma $= 0.1$ and step size $= 10$)
Weight decay	0.0001
Max epochs	100 (with early stopping after 10)

Bias Mitigation Models. We adapt implementations of existing bias mitigation methods from the github repository MEDFAIR, using the same backbone and core parameters as in Table 1. We select methods which encompass different types of bias mitigation approaches and which had good results in the MEDFAIR benchmark [33] and try to mitigate age-, assessment-centre-, and sex-related disparities.

We test **Resampling** of minority subgroups as a pre-processing method [12]. In addition, we explore a range of in-processing methods including **Group Distributionally Robust Optimisation (GroupDRO)** which minimises worst-group loss [16,20], **Orthogonally Disentangled Representations (ODR)**, which disentangles the representations of subgroup-related features and task-relevant features [22], **Domain-Independent learning (DomainInd)** where each subgroup has its own final classification layer [29], and **Learning-Not-to-Learn (LNL)**, an adversarial learning method [15]. We also implement **Stochastic Weight Averaging Densely (SWAD)** [7] which is a general robustness method (and therefore does not require subgroup information) and pair it with resampling (**ReSWAD**). Finally, we implement a post-processing method (not in MEDFAIR), **Recalibration**, where a different decision threshold is calculated for each subgroup. We train all models three times with different random seeds on NVIDIA A100 GPU's.

Model Evaluation. Model evaluation is based upon the mean Receiver Operating Characteristic Area Under the Curve (AUC), accuracy, precision, and recall scores for the three runs of each model. We consider overall performance and performance across different subgroups (both minimum performance and best- and worst- performance gap). All code is available at https://github.com/anissa218/ MEDFAIR_UKBB.

3   Results and Discussion

Performance and Disparities of the Baseline Model. The baseline InceptionV3 model achieves $73 \pm 0.01\%$ accuracy and $71 \pm 0.00\%$ AUC in hypertension classification, with precision and recall values of $81 \pm 0.04\%$ and $83 \pm 0.01\%$, respectively. However, a more granular assessment reveals significant disparities across certain subgroups (Table A1). For instance, as shown in Fig. 1, the model's AUC varies by over 15% between different age groups and 10% between centres, with the worst-group AUC being substantially lower than the average AUC of 0.71. Some subgroups also exhibit substantial differences in recall (which would translate to underdiagnosis) of 10 to 32%, including different age groups, assessment centres, alcohol consumers, and ethnic groups.

Origins of Performance Disparities. Next, we aim to understand why these disparities appear. We investigate whether they can be attributed to varying underlying characteristics across subgroups, such as differences in age or sex

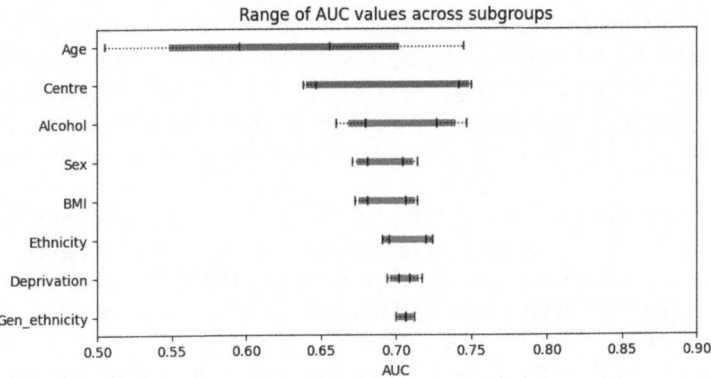

Fig. 1. For some subgroupings, the baseline model shows large disparities in test set AUC between worst- and best- performing subgroups, far below and above the average AUC of 0.71. Error bars represent standard deviation across the three random seeds.

distribution. However, regardless of the attribute we condition on, the worst-performing assessment centre, centre f, shows much lower AUC (results on age conditioning are shown in Table A2). Such trends are also preserved for sex- and age-related disparities. Additionally, we use the Automorph pipeline [31] to assess the quality of all images and use this as a conditioning variable. We find that image quality does not explain these disparities either. We also consider shifts in prevalence, as correlation between an attribute and the target label can cause bias [13]. This is evident in age- and sex-related disparities, where hypertension shows a strong positive correlation with age (Figure A1), and men have a higher prevalence of hypertension. However, this does not explain centre disparities, as the worst-performing subgroup has approximately 76% images with hypertension, which falls within the range of other centres (69%–80%).

Further, these differences cannot simply be attributed to data imbalance. For centre and sex-related disparities, all groups are evenly represented. However, for age-related disparities, data imbalance may play a role. The oldest age group, which has the lowest AUC, is also underrepresented, comprising only 2.5% of the images.

Another emerging hypothesis in fair ML research is that disparities arise due to unequal model generalisation across subgroups. Despite uniform and strong performance on training data, generalisation differences on unseen data can emerge [11,20]. As shown in Table 2, there is a noticeable decrease in worst-group AUC relative to the decrease in overall AUC between training data and test data for different centres. Similarly, the gap between centres increases on unseen data, suggesting that the model's generalisation varies across these centres. The difference is not as striking for age and sex subgroups, and most likely simply linked to overall performance decrease on unseen data. We further investigate whether there is a shift in generalisation during training; a point where the model starts overfitting to certain subgroups but not others (and thus increasing the gap between subgroups) as identified in [11]. However our analyses do not reveal any evidence of a specific point where this could occur (Figures A2).

Table 2. Age, centre, and sex disparities across seen and unseen data (Test AUC - Train AUC). While disparities increase in unseen test data for all groups, the increase is strongest for assessment centres, suggesting unequal generalisation. Standard deviation of the three random seeds shown in parentheses.

Subgroup	Δ Overall AUC	Δ Min AUC	Δ AUC Gap
Age	−0.031 (0.011)	−0.037 (0.055)	0.004 (0.044)
Centre	−0.031 (0.011)	−0.045 (0.014)	0.032 (0.006)
Sex	−0.031 (0.011)	−0.034 (0.011)	0.009 (0.002)

Finally, we investigate whether the model's learnt representations can provide insight on subgroup disparities. We analyse each image in the model's penultimate layer feature space through a 4-component principal component analysis (which explains over 85% of the variance). As expected, we find strong separation between the projected features of images with and without hypertension, and consequently between images of different age groups due to their strong correlation. However, we also observe an unexpected outlier from the distribution of images from the worst-performing centre (f). There is a clear difference in the kernel density estimates of some principal components from this centre

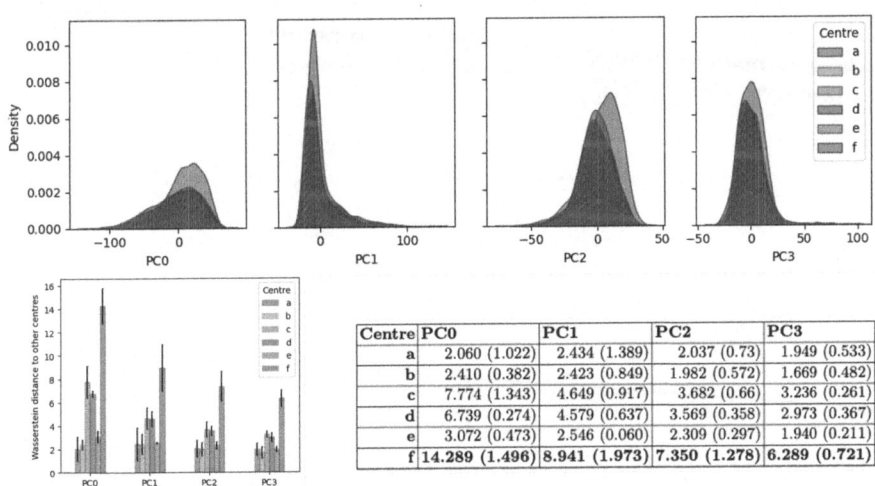

Fig. 2. Kernel density estimation of the first 4 principal components (PC) of the features extracted from the baseline model's penultimate layer grouped by centre. Table of mean Wasserstein distance of features between one centre and the other 5 for the 3 random seeds. f's feature distribution is clearly an outlier across some PCs.

and a consistently increased Wasserstein distance separating the distribution of features from centre f to the other centres (Fig. 2). Although this does not prove this information is being used for predictions, it is noteworthy that such a shift exists, one that cannot be explained by any of the other available variables.

Overall Performance of Mitigation Models. We then train a number of bias mitigation methods with the objective of reducing the most significant disparities: age, assessment centre, and sex. Initially, we assess how these methods impact overall model performance across all samples, examining whether "levelling down" occurs [32]. Regarding age mitigation, SWAD is the only method capable of maintaining overall AUC, whereas all other mitigation methods result in a decrease in AUC, particularly gDRO (Fig. 3). Interestingly, this decrease in AUC is less pronounced in the assessment centre mitigation models. Only LNL and ODR show a notable decrease in AUC and precision, whereas the other models show similar overall performance across all four metrics (Figure A3). Sex disparity mitigation has a more variable effect (see Figure A4).

Disparity Reduction. Overall, no methods achieve their intended effects of reducing disparities and boosting worst group performance. For age-related disparities, DomainInd is the only model which shows some effectiveness; it decreases accuracy, AUC, and recall gap relative to baseline while also increasing worst-group performance (Table 3). However, it also causes a slight reduction in overall performance (Figure A3). SWAD performance is generally similar to baseline performance, but other models decrease min AUC and min precision.

For centre-related disparities, the effectiveness of the models in improving fairness is very limited, especially in boosting worst-group performance. SWAD is the only method which maintains or slightly improves upon baseline disparities (Table 3). Other methods have negative effects on at least one of the met-

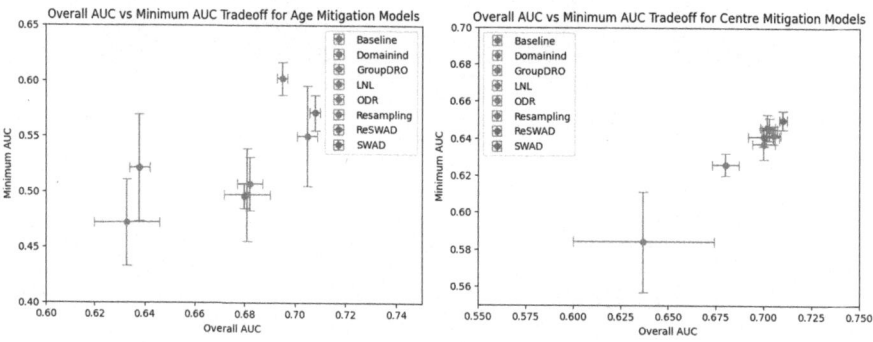

Fig. 3. Overall AUC of age mitigation models (left) and centre mitigation models (right) relative to worst-group AUC. Most models worsen both overall and minimum performance relative to the baseline (red point), especially for age mitigation. Error bars represent standard deviation for 3 random seeds. (Color figure online)

rics. For instance, resampling increases accuracy gap, ODR lowers min AUC by 0.02, and recalibration lowers min recall by 0.02. We also note that the optimal per-subgroup decision thresholds (for recalibration) range from 0.50 to 0.73, suggesting the baseline model does not uniformly adapt to the characteristics of different subgroups.

Table 3. Performance disparities across age groups and assessment centres for their respective mitigation models. DomainInd is the only method able to reduce most age disparities relative to the baseline model, while no models are able to consistently reduce assessment centre disparities. Standard deviation of the three random seeds shown in parentheses.

	Model	Acc. Gap↓	Min Acc.↑	AUC Gap↓	Min AUC↑	Prec. Gap↓	Min Prec.↑	Rec. Gap↓	Min Rec.↑
Age	Baseline	0.187 (0.017)	0.639 (0.017)	0.15 (0.04)	0.55 (0.045)	0.129 (0.005)	0.724 (0.008)	0.328 (0.045)	0.643 (0.055)
	DomainInd	**0.158 (0.018)**	0.658 (0.007)	**0.101 (0.013)**	**0.602 (0.015)**	0.161 (0.005)	0.702 (0.004)	**0.189 (0.029)**	0.743 (0.023)
	GroupDRO	0.245 (0.014)	0.6 (0.014)	0.147 (0.018)	0.472 (0.039)	0.178 (0.012)	0.668 (0.012)	0.341 (0.019)	0.659 (0.019)
	LNL	0.236 (0.01)	0.61 (0.01)	0.107 (0.053)	0.521 (0.048)	0.163 (0.008)	0.684 (0.009)	0.349 (0.054)	0.649 (0.055)
	ODR	0.184 (0.021)	0.647 (0.008)	0.174 (0.038)	0.507 (0.024)	0.142 (0.004)	0.705 (0.006)	0.277 (0.026)	0.704 (0.011)
	Recalibration	0.175 (0.007)	**0.664 (0.004)**	0.15 (0.04)	0.55 (0.045)	0.156 (0.007)	0.699 (0.005)	0.22 (0.024)	**0.768 (0.013)**
	Resampling	0.181 (0.006)	0.647 (0.014)	0.185 (0.053)	0.497 (0.042)	0.136 (0.016)	0.71 (0.017)	0.283 (0.014)	0.692 (0.013)
	ReSWAD	0.187 (0.027)	0.647 (0.018)	0.191 (0.013)	0.496 (0.011)	0.122 (0.007)	0.725 (0.006)	0.323 (0.056)	0.66 (0.048)
	SWAD	0.192 (0.003)	0.646 (0.003)	0.13 (0.018)	0.571 (0.016)	**0.121 (0.005)**	**0.729 (0.009)**	0.339 (0.021)	0.65 (0.024)
Centre	Baseline	0.061 (0.012)	0.706 (0.013)	0.104 (0.004)	0.642 (0.005)	0.097 (0.012)	0.776 (0.01)	0.149 (0.013)	0.775 (0.029)
	DomainInd	**0.055 (0.019)**	0.712 (0.004)	0.106 (0.005)	0.646 (0.007)	0.111 (0.007)	0.763 (0.002)	0.189 (0.027)	0.78 (0.034)
	GroupDRO	0.061 (0.006)	0.71 (0.001)	0.105 (0.003)	0.637 (0.008)	0.106 (0.002)	0.765 (0.003)	0.183 (0.009)	0.779 (0.014)
	LNL	0.082 (0.007)	0.682 (0.015)	**0.089 (0.017)**	0.584 (0.027)	**0.092 (0.004)**	0.751 (0.01)	0.197 (0.071)	0.785 (0.088)
	ODR	0.065 (0.017)	0.71 (0.006)	0.097 (0.003)	0.626 (0.006)	0.098 (0.006)	0.765 (0.006)	**0.137 (0.023)**	**0.808 (0.015)**
	Recalibration	0.118 (0.022)	0.711 (0.015)	0.104 (0.004)	0.642 (0.005)	0.079 (0.011)	**0.781 (0.007)**	0.19 (0.071)	0.755 (0.05)
	Resampling	0.085 (0.013)	0.712 (0.011)	0.098 (0.013)	0.645 (0.006)	0.101 (0.006)	0.769 (0.003)	0.146 (0.022)	0.799 (0.033)
	ReSWAD	0.082 (0.008)	0.712 (0.008)	0.097 (0.005)	0.641 (0.005)	0.107 (0.005)	0.762 (0.006)	0.174 (0.019)	0.793 (0.012)
	SWAD	0.06 (0.019)	**0.715 (0.013)**	0.095 (0.009)	**0.65 (0.005)**	0.102 (0.007)	0.772 (0.009)	0.156 (0.016)	0.776 (0.046)

4   Conclusions

Our model trained with retinal images from the UKBB shows notably poor performance on certain subgroups of the population. In particular, although some level of age- or sex-related disparities could be expected due to differences in biological manifestation or prevalence of hypertension, centre disparities (which cannot be explained by any of the investigated confounders), are unexpected given the standardisation of the UKBB. These disparities would lead to unfair outcomes if such a model was deployed. This highlights the importance of systematically conducting a granular assessment of a model's performance.

Moreover, existing methods largely fail to mitigate these disparities. Most methods, particularly for age disparity mitigation, have a detrimental effect on overall performance. Even worse, few really improve fairness, and while some may show marginal improvement in one scenario, they adversely impact others. For instance, the DomainInd model slightly improves age- and sex-related disparities but does not show improvements in assessment-centre disparities. No method is actually able to boost performance for assessment centre f, suggesting that further methodological advancements are necessary, or that perhaps a maximum performance has already been reached rendering mitigation efforts ineffective. These observations highlight how applying bias mitigation methods indiscriminately may actually worsen overall outcomes and exacerbate existing disparities, concordant with recent findings in MEDFAIR [33]. Overall, it appears important to precisely characterise biases and their underlying causes, as this understanding is crucial for informing appropriate mitigation strategies.

Future work should continue to develop a framework to better characterise disparities, for example building off previous work done in [13, 14]. We consider a very narrow scenario of hypertension prediction from retinal images, but it would be interesting to see how these findings extend to other retinal image tasks and other image modalities. It would also be of interest to conduct a more in-depth exploration of the UKBB dataset specifically, in order to understand the interplay between selection bias, dataset standardisation, and subsequent model biases, and shed light on why some assessment centres showed such disparate performance. Investigations of this kind are increasingly important given the rise in large databases and initiatives like the UKBB, and the need to ensure downstream findings stay as unbiased as possible.

Acknowledgments. This research has been conducted using data from UK Biobank, a major biomedical database, with access provided through application 80521. This work was supported by the EPSRC grant number EP/S024093/1 and the Centre for Doctoral Training in Sustainable Approaches to Biomedical Science: Responsible and Reproducible Research (SABS: R3) Doctoral Training Centre, University of Oxford and by GE Healthcare. The computational aspects of this research were supported by the Wellcome Trust Core Award Grant Number 203141/Z/16/Z and the NIHR Oxford BRC. The views expressed are those of the author(s) and not necessarily those of the NHS, the NIHR or the Department of Health.

Disclosure of Interests. None.

References

1. Resource 100237: Optical-coherence tomography procedures using ACE. https://biobank.ndph.ox.ac.uk/showcase/refer.cgi?id=100237
2. Allen, N.E., Lacey, B., Lawlor, D.A., et al.: Prospective study design and data analysis in UK Biobank **16**(729), eadf4428 (2024). https://doi.org/10.1126/scitranslmed.adf4428
3. Bevan, P., Atapour-Abarghouei, A.: Skin deep unlearning: artefact and instrument debiasing in the context of melanoma classification. In: Proceedings of the 39th International Conference on Machine Learning, pp. 1874–1892. PMLR (2022)
4. Bradley, V., Nichols, T.E.: Addressing selection bias in the UK biobank neurological imaging cohort (2022). https://doi.org/10.1101/2022.01.13.22269266. https://www.medrxiv.org/content/early/2022/01/24/2022.01.13.22269266
5. Burlina, P., Joshi, N., Paul, W., et al.: Addressing artificial intelligence bias in retinal diagnostics **10**(2), 13 (2021). https://doi.org/10.1167/tvst.10.2.13
6. Castelnovo, A., Crupi, R., Greco, G., et al.: A clarification of the nuances in the fairness metrics landscape **12**(1), 4209 (2022). https://doi.org/10.1038/s41598-022-07939-1

7. Cha, J., Chun, S., Lee, K., et al.: SWAD: domain generalization by seeking flat minima. In: Advances in Neural Information Processing Systems, vol. 34, pp. 22405–22418. Curran Associates, Inc. (2021)

8. Coyner, A.S., Singh, P., Brown, J.M., et al.: Association of biomarker-based artificial intelligence with risk of racial bias in retinal images **141**(6), 543–552 (2023). https://doi.org/10.1001/jamaophthalmol.2023.1310

9. Diana, E., Gill, W., Kearns, M., et al.: Convergent algorithms for (relaxed) minimax fairness. CoRR abs/2011.03108 (2020). https://arxiv.org/abs/2011.03108

10. Du, Y., Xue, Y., Dharmakumar, R., et al.: Unveiling fairness biases in deep learning-based brain MRI reconstruction. In: Wesarg, S., et al. (eds.) CLIP EPIMI FAIMI 2023. LNCS, vol. 14242, pp. 102–111. Springer, Cham (2023). https://doi.org/10.1007/978-3-031-45249-9_10

11. Dutt, R., Bohdal, O., Tsaftaris, S.A., et al.: FairTune: optimizing parameter efficient fine tuning for fairness in medical image analysis (2024). https://arxiv.org/abs/2310.05055

12. Idrissi, B.Y., Arjovsky, M., Pezeshki, M., et al.: Simple data balancing achieves competitive worst-group-accuracy. In: Proceedings of the First Conference on Causal Learning and Reasoning, pp. 336–351. PMLR (2022)

13. Jones, C., Castro, D.C., De Sousa Ribeiro, F., et al.: A causal perspective on dataset bias in machine learning for medical imaging **6**(2), 138–146 (2024). https://doi.org/10.1038/s42256-024-00797-8

14. Jones, C., Roschewitz, M., Glocker, B.: The role of subgroup separability in group-fair medical image classification (2023). https://arxiv.org/abs/2307.02791

15. Kim, B., Kim, H., Kim, K., et al.: Learning not to learn: training deep neural networks with biased data. In: 2019 IEEE/CVF Conference on Computer Vision and Pattern Recognition (CVPR), pp. 9004–9012. IEEE (2019). https://doi.org/10.1109/CVPR.2019.00922

16. Kumar, N., Shrestha, R., Li, Z., et al.: Distributionally robust optimization and invariant representation learning for addressing subgroup underrepresentation: mechanisms and limitations. In: Wesarg, S., et al. (eds.) CLIP EPIMI FAIMI 2023. LNCS, vol. 14242, pp. 183–193. Springer, Cham (2023). https://doi.org/10.1007/978-3-031-45249-9_18

17. Lyall, D.M., Quinn, T., Lyall, L.M., et al.: Quantifying bias in psychological and physical health in the UK Biobank imaging sub-sample **4**(3), fcac119 (2022). https://doi.org/10.1093/braincomms/fcac119

18. Poplin, R., Varadarajan, A.V., Blumer, K., et al.: Prediction of cardiovascular risk factors from retinal fundus photographs via deep learning **2**(3), 158–164 (2018). https://doi.org/10.1038/s41551-018-0195-0

19. Puyol-Antón, E., et al.: Fairness in cardiac MR image analysis: an investigation of bias due to data imbalance in deep learning based segmentation. In: de Bruijne, M., et al. (eds.) MICCAI 2021. LNCS, vol. 12903, pp. 413–423. Springer, Cham (2021). https://doi.org/10.1007/978-3-030-87199-4_39

20. Sagawa, S., Koh, P.W., Hashimoto, T.B., et al.: Distributionally robust neural networks for group shifts: on the importance of regularization for worst-case generalization (2020). https://doi.org/10.48550/arXiv.1911.08731

21. Sanchez, P., Voisey, J.P., Xia, T., et al.: Causal machine learning for healthcare and precision medicine **9**(8), 220638 (2022). https://doi.org/10.1098/rsos.220638

22. Sarhan, M.H., Navab, N., Eslami, A., Albarqouni, S.: Fairness by learning orthogonal disentangled representations. In: Vedaldi, A., Bischof, H., Brox, T., Frahm, J.-M. (eds.) ECCV 2020. LNCS, vol. 12374, pp. 746–761. Springer, Cham (2020). https://doi.org/10.1007/978-3-030-58526-6_44

23. Schoeler, T., Speed, D., Porcu, E., et al.: Participation bias in the UK biobank distorts genetic associations and downstream analyses **7**(7), 1216–1227 (2023). https://doi.org/10.1038/s41562-023-01579-9
24. Seyyed-Kalantari, L., Zhang, H., McDermott, M.B.A., et al.: Underdiagnosis bias of artificial intelligence algorithms applied to chest radiographs in under-served patient populations **27**(12), 2176–2182 (2021). https://doi.org/10.1038/s41591-021-01595-0
25. Sudlow, C., Gallacher, J., Allen, N., et al.: UK biobank: an open access resource for identifying the causes of a wide range of complex diseases of middle and old age **12**(3), e1001779 (2015). https://doi.org/10.1371/journal.pmed.1001779
26. Swanson, J.M.: The UK Biobank and selection bias **380**(9837), 110 (2012). https://doi.org/10.1016/S0140-6736(12)61179-9
27. Szegedy, C., Vanhoucke, V., Ioffe, S., et al.: Rethinking the inception architecture for computer vision (2015). https://arxiv.org/abs/1512.00567
28. Tobin, M.D., Sheehan, N.A., Scurrah, K.J., et al.: Adjusting for treatment effects in studies of quantitative traits: antihypertensive therapy and systolic blood pressure **24**(19), 2911–2935 (2005). https://doi.org/10.1002/sim.2165
29. Wang, Z., Qinami, K., Karakozis, I.C., et al.: Towards fairness in visual recognition: effective strategies for bias mitigation (2020)
30. Whelton, P.K., Carey, R.M., Aronow, W.S., et al.: ACC guideline for the prevention. Detect. Eval. Manag. High Blood Pressure Adults **71**(6), 1269–1324 (2017). https://doi.org/10.1161/HYP.0000000000000066
31. Zhou, Y., Wagner, S.K., Chia, M.A., et al.: AutoMorph: automated retinal vascular morphology quantification via a deep learning pipeline. Transl. Vis. Sci. Technol. **11**(7), 12 (2022). https://doi.org/10.1167/tvst.11.7.12
32. Zietlow, D., Lohaus, M., Balakrishnan, G., et al.: Leveling down in computer vision: pareto inefficiencies in fair deep classifiers (2022). https://arxiv.org/abs/2203.04913
33. Zong, Y., Yang, Y., Hospedales, T.: MEDFAIR: benchmarking fairness for medical imaging. In: International Conference on Learning Representations (ICLR) (2023)

Investigating Gender Bias in Lymph-Node Segmentation with Anatomical Priors

Ricardo Coimbra Brioso[1(✉)], Damiano Dei[2,3], Nicola Lambri[2,3],
Pietro Mancosu[2,3], Marta Scorsetti[2,3], and Daniele Loiacono[1]

[1] Department of Electronics, Information and Bioengineering, Politecnico di Milano,
Milan, Italy
ricardo.brioso@polimi.it
[2] Department of Biomedical Sciences, Humanitas University, Milan, Italy
[3] Radiotherapy and Radiosurgery, IRCCS Humanitas Research Hospital,
Rozzano, Italy

Abstract. Radiotherapy requires precise segmentation of organs at risk
(OARs) and of the Clinical Target Volume (CTV) to maximize treatment
efficacy and minimize toxicity. While deep learning (DL) has significantly
advanced automatic contouring, complex targets like CTVs remain challenging. This study explores the use of simpler, well-segmented structures
(e.g., OARs) as Anatomical Prior (AP) information to improve CTV segmentation. We investigate gender bias in segmentation models and the
mitigation effect of the prior information. Findings indicate that incorporating prior knowledge with the discussed strategies enhances segmentation quality in female patients and reduces gender bias, particularly
in the abdomen region. This research provides a comparative analysis
of new encoding strategies and highlights the potential of using AP to
achieve fairer segmentation outcomes.

Keywords: clinical target volume · CTV · lymph nodes · TMI ·
TMLI · semantic segmentation · deep learning · fairness ·
visualization · anatomical prior

1 Introduction

For several types of cancer, radiotherapy is the most effective treatment, and it is
used in more than 50% of cancer patients as the main treatment, concurrent, or
perioperative multimodal treatments [1]. Radiotherapy uses ionizing radiation
to kill cancer cells and shrink tumors. To avoid toxicity and side effects, the
radiation dose must be delivered to the tumor while sparing the surrounding
healthy tissues. Accordingly, radiation oncologists (ROs) define the OARs and
the CTV which is the volume that corresponds to the tissue to be irradiated
and includes a small margin that accounts for body changes and movement.
The ROs dedicate several hours to this contouring in the full-body Computed
Tomography (CT) and the automation of these segmentations would speed up
this process.

E. Puyol-Antón et al. (Eds.): FAIMI 2024/EPIMI 2024, LNCS 15198, pp. 151–160, 2025.
https://doi.org/10.1007/978-3-031-72787-0_15

Advancements in DL proved to be a very effective and consistent tool for automating the image contouring process in radiotherapy [6]. Healthcare centers and hospitals are beginning to use tools embedded with DL models to automate contouring. This is possible due to research in DL, hardware improvements, and, data availability. Medical imaging tasks are characterized by having a smaller amount of data in comparison to other computer vision tasks. This coupled with the complexity of some targets, such as CTV in Total Marrow and Lymph node Irradiation (TMLI) [19], increases the difficulty of improving automatic segmentation in this particular task.

Segmentation models can perform better based on architectural changes in the network, this has motivated iterative improvements in models' performance. In this article, it is studied an alternative path for improving the segmentation quality. Available segmented structures that are simpler than the CTV (e.g.: OARs) are available in more datasets and can be obtained with open-source models [18]. Using OARs and other structures as prior information to the segmentation model is still unexplored. This work provides the first comparative analysis of strategies to encode prior knowledge.

While assessing the models, gender disparity is found in the segmentation performance. Gender bias has always been present in many DL tasks but research in this field is limited despite its importance. Approaches to mitigate gender bias in segmentation models are even less common [2]. To accurately measure fairness, different metrics are applied and the segmentation results are observed in different regions of the patient. The findings of this work move us towards exploring Anatomical Priors (AP) to mitigate gender bias. Using already available data that contains gender-specific information improves the model's performance in the abdomen regions. Even structures that exist in both genders can have information about the gender, for example, the size and ratio of the structure when compared to other structures can vary between female and male. The main contributions of this work are: (i) exploring and comparing new encoding strategies for AP in the segmentation model's training. (ii) analyzing gender bias and the effects of AP in mitigating it in the different regions of the body.

2   Related Work

Radiotherapy has automatic segmentation solutions for OARs and CTV, although CTV remains rarer to see segmented. For OARs and other structures, we can see works for different regions: the head and neck [10], thorax [20], abdomen [17], and pelvis [12].

Cervical cancer CTV segmentation in [11] is obtained using DpnUNet, a similar architecture to the U-Net [14], it combines Residual and Dense blocks in the encoder, enhancing the ability to recover and refine abstract features. It also adds three adjacent slices as three new channels in the input of the network, making it a 2.5D architecture. In [16], segmentation models for rectal and cervical cancer CTV are developed using U-Net-based architectures or DeeplabV3+ [5]. The work of [16] concludes that time is saved by using the CTV automatic contouring, even when the latter needs corrections. In [12], an automatic CTV contouring is compared with an RO's contouring and observes that automatic contours are on par with manual contouring and save RO's time.

The CTV is a complex agglomerate of structures that is hard to segment, in some works it is shown that adding anatomical prior information about the segmentation target and incorporating it in the loss function can improve its performance and anatomical plausibility [3].

In [9], several single-organ models are used to learn anatomical invariance across different subjects and datasets and using this information as priors for other segmentation models, thus, improving performance. To focus the attention of the network in the pancreatic area, in [15], a mask indicating the region of the pancreas is added as a prior.

Following the CTV segmentation experiments with AP, it was analyzed that this addition mitigated the gender bias present in the segmentation performance. Analyzing biases in deep learning models is an important topic that has only gained visibility in recent years [4]. In medical imaging segmentation, only a handful of articles investigate model biases. In [2], several training approaches were developed that helped to mitigate racial bias. For example, a stratified batch to ensure a balance between racial groups in each iteration of the training or an additional DL classification network that classified the gender before performing the training segmentation.

The work [8] compared gender and race bias in three DL segmentation architectures and a transformer architecture in cardiac MR segmentation. The models' biases differ for each gender or race and respond differently to the percentage of minority group's data present in the training set of the model.

Anatomical priors and gender analysis works for segmentation models are scarce but very important due to their applicability and consequences in the real world. For the first time, diverse ways to encode AP information are compared, while analyzing its effects on gender bias of segmentation models.

3   Methods

3.1   Data

The data used contains 45 full-body CT 3D volumes, 25 male and 19 female patients. A free-breathing, non-contrast CT scan with a 5-mm slice thickness was acquired for each patient using a BigBore CT system (Philips Healthcare, Best, Netherlands) [13] and each axial slice of the CT has a resolution of 512×512 pixels. The patients are candidates to undergo radiotherapy treatment and for every CT volume, several structures were delineated by a RO, including the Clinical Target Volume (CTV) that corresponds to the goal target of this segmentation task. For two steps of the experimental design, structures generated with the help of TotalSegmentator were used. Firstly, several structures from TotalSegmentator were used as AP to train the segmentation models. Secondly, the vertebra T1, the stomach, and, vertebra L4. were used to separate each patient into four regions: the Head and Neck (HN), the thoracic (THX), the abdomen (ABDM), and the pelvic region (PELV). The separation of the regions is used to evaluate the performance of the CTV segmentation in more detail in each region.

3.2   Anatomical Prior Strategies

Several experiments were made by inputting anatomical prior information (segmentation of other structures or OARs). The APs were inputted in different ways: Multiple Intensity Z-Score (MI-Z): The structures (spleen, liver, eyes, kidneys, femurs, stomach, heart) present in the ground truth (GT) of the dataset were added as an additional channel to the input after encoding each structure with a different value of intensity (from 0 to 255) with Z-Score normalization on the APs. Equal Intensity Z-Score (EQ-Z): Similar to the previous inputting but the value of intensity is the same for every structure (intensity of 255). Cropped CT Z-Score (Crop-Z): Two additional channels were used in this inputting strategy, both of them included the original intensities of the CT image in the image's position of the additional structures, while the image's external part to these structures had an intensity value of 0 in one channel and 255 in the other channel. Multiple Intensity (MI): Identical to the MI-Z inputting but without Z-Score normalization. Multiple Intensity with TotalSegmentator (MI-TS): The structures used in this inputting technique were not from the GT but from TotalSegmentator and no normalization was applied. The structures used in this case were: humeri, scapulae, clavicles, femurs, hips, sacrum, spleen, liver, stomach, urinary bladder, pancreas, kidneys, and, iliopsoas muscles. The structures were encoded in the same way as the first strategy, each with its intensity and without Z-Score normalization on the APs.

Providing contextual information from adjacent organs could improve segmentation performance and reduce sex-based bias.

3.3   Training and Evaluation

We applied the nnU-Net [7] framework, an adapting algorithm that analyzes the dataset's characteristics, such as resolution and pixel spacing. This information is used for pre-processing and tuning the training parameters of the nnU-Net. The evaluation metrics used were the Dice Score (DSC) and Hausdorff Distance (HD).

The $DSC = \frac{2|X \cap Y|}{|X|+|Y|}$ measures the overlap between the GT and the predicted mask, with X representing the set of positive pixels in the GT and Y representing the set of positive pixels in the prediction.

The $HD = max\{\max_{x \in S_X} d(x, S_Y), \max_{y \in S_Y} d(y, S_X)\}$ measures the maximum distance of all the nearest distances between the surfaces of the two sets X and Y, denoted as S_X and S_Y. To mitigate the impact of outliers on HD values, we employed HD95, which excludes the top 5% highest HD values.

An evaluation focused on different regions of the patient was conducted to localize the model's biases. To evaluate the performance of the CTV segmentation model in different regions of the patient, structures segmented by TotalSegmentator were used to locate the boundaries of four regions of the patient's body:

HN, THX, ABDM, PELV. Dividing the patient analysis is useful to see if the bias is related to a specific part of the patient that is harder to segment. Several metrics were developed to understand the underlying patterns and location of the gender bias:

- **Average Gender Difference (AGD)**: The average difference in Dice scores between male and female patients.
- **Median Gender Difference (MGD)**: Measures the median difference in Dice scores between male and female patients.
- **Quartile Difference (QD)**: Calculated as the maximum of the difference between the third quartile of male DSC and the first quartile of female DSC, and the difference between the third quartile of female DSC and the first quartile of male DSC.

Due to the quantity of available data, a statistical analysis is not presented.

4   Results and Discussion

In Table 1, we can see the whole-body performance of the different AP models and their respective gender biases. All models have similar median DSC values, ranging from 82.98% to 83.46%, indicating comparable segmentation performance across models. The HD95 median across all models is around 4.8 mm to 4.9 mm, suggesting minimal changes on the surface of the contouring with AP. The MI-TS model has the highest median DSC at 83.46%, suggesting a slight edge in segmentation performance.

The Base model shows a median DSC difference of 5.25% between females (79.61%) and males (84.86%), indicating gender bias. There is a gender median DSC disparity in every model, varying from 3.82% to 5.33%. MI-Z improved DSC performance for females (80.58%) compared to the base model, and a slightly lower performance for males (84.40%), having one of the smallest median DSC differences between males and females across all models. This indicates a reduction in gender disparity. The MI model improved median DSC while maintaining a high male median DSC. Models MI-Z, EI-Z, Crop-Z, MI, and MI-TS show a trend toward reducing gender bias, improving the DSC for female patients without significantly compromising the performance of male patients.

Table 1. Different Multiple-Input Models and their Gender Bias

Model	DSC Med.	HD95 Med.	DSC F-Med.	DSC M-Med.
Base	83.40%	4.83	79.61%	84.86%
MI-Z	83.36%	4.76	80.58%	84.40%
EI-Z	83.26%	4.83	80.35%	84.32%
Crop-Z	83.14%	4.76	80.50%	84.59%
MI	82.98%	4.91	80.46%	85.33%
MI-TS	83.46%	4.69	80.24%	84.20%

Table 2 shows that the Base model has a high median DSC across all regions, with values ranging from 82.63% (PELV) to 85.31% (HN). There is a notable gender disparity, with males having a higher median DSC than females in all regions. In the HN region, males achieve a median DSC of 85.90% compared to 82.90% for females. The MI-Z model improves the median DSC in the ABDM region to 86.00%, indicating an enhancement in segmentation performance in this region while showing a decrease in the HN (84.84%) and THX (82.97%).

The EI-Z model maintains consistent performance across regions, with slight improvements in the THX (83.12%) and ABDM (85.76%) regions. The Crop-Z had the greatest improvement for the HN region in general and for male patients while the MI-TS improved the HN performance for female patients (↑ 0.3%).

The MI-TS model shows an improvement in the THX region with a median DSC of 83.10%. Performance in other regions is slightly lower compared to the Base and MI models. Gender disparities are observed, with males consistently achieving higher median DSCs.

The region that improved the most was the ABDM region, where all the AP models improved DSC for female patients. This outcome is related to the high number of simultaneous AP structures in this region, as seen in Fig. 1. In Fig. 1, the first two columns correspond to an example where having APs of the heart, the liver, and the stomach, improves the segmentation of the lymph nodes around the stomach (lymph nodes are part of the CTV). The second example of another patient (located in the last two columns) shows that the AP of the stomach improved the delineation of the surrounding lymph nodes once more.

Table 2. Different Multiple-Input Models and their Gender Bias in different regions.

Model	Med. DSC ↑	HN	THX	ABDM	PELV
Base	Total	85.31%	**83.58%**	84.34%	**82.63%**
	Female	82.90%	79.90%	82.50%	**81.20%**
	Male	85.90%	85.20%	86.30%	83.10%
MI-Z	Total	84.84%	82.97%	**86.00%**	82.36%
	Female	79.50%	**80.40%**	**85.80%**	80.80%
	Male	86.50%	85.80%	86.30%	83.00%
EI-Z	Total	84.63%	83.12%	85.76%	82.30%
	Female	79.90%	79.90%	85.70%	80.90%
	Male	86.10%	85.40%	86.30%	83.20%
Crop-Z	Total	85.35%	82.86%	85.58%	82.34%
	Female	79.40%	80.20%	85.00%	80.50%
	Male	86.30%	85.50%	85.90%	82.90%
MI	Total	85.92%	82.59%	84.62%	82.50%
	Female	83.10%	80.10%	84.30%	80.80%
	Male	**86.90%**	85.80%	84.90%	82.80%
MI-TS	Total	84.97%	83.10%	84.50%	82.05%
	Female	**83.20%**	80.00%	83.20%	81.00%
	Male	86.30%	85.20%	86.40%	82.50%

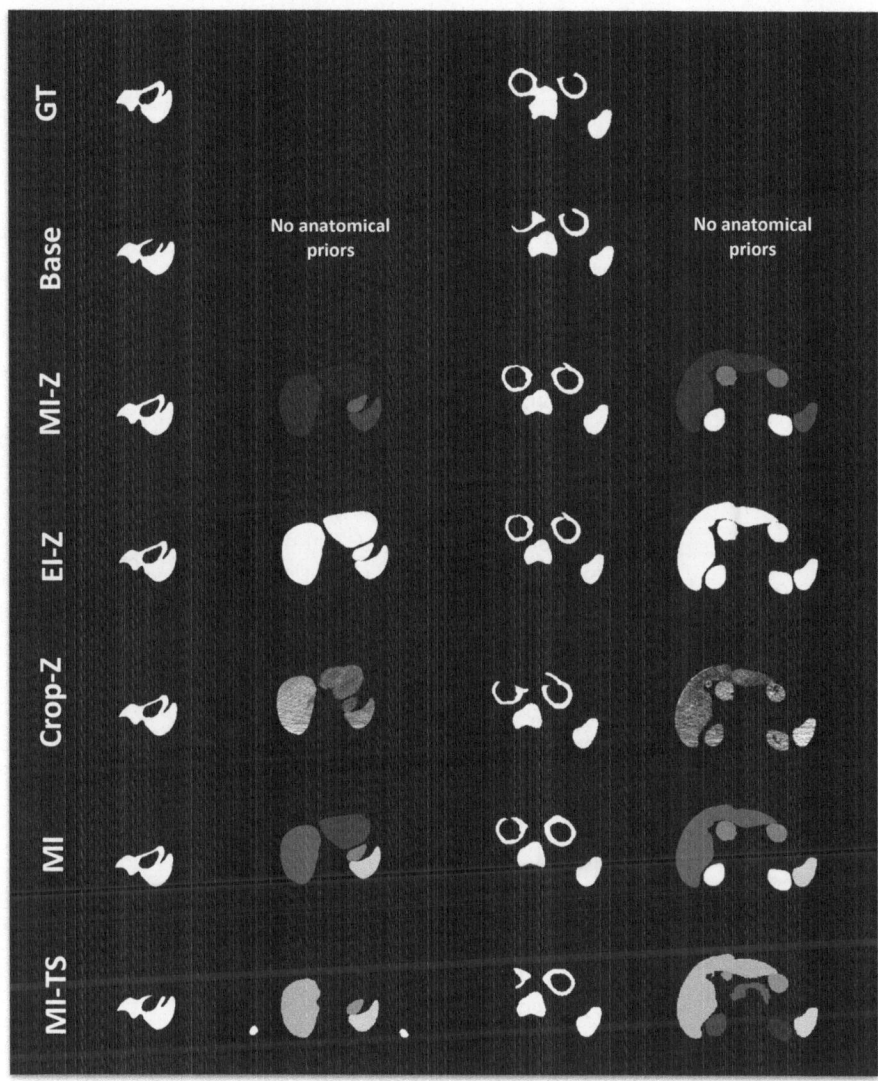

Fig. 1. Examples of improvements in the CTV segmentation predictions due to anatomical prior addition in the ABDM region.

Table 3 shows three fairness metrics: AGD, MGD, and, QD for the AP models.

The MI-TS model shows the lowest AGD and MGD for HN, THX, and PELV. The regions with higher disparity are the HN, THX, and ABDM. The PELV region contains the majority of gender-specific structures but has the lowest disparity, this could indicate that the CTV in this region is easier to segment due to its relatively lesser area in this region and that the APs used were insufficient to help the model improve its performance. Despite some models reducing AGD and MGD, the regions remain with high QD favoring males, reflecting bias. The insights provided by fairness metrics are crucial for selecting and improving segmentation models and ensuring fairness and similar accuracy across different patient demographics and real-life applications.

Table 3. Different Multiple-Input Models and their Gender Bias in different regions for the developed fairness metrics.

Model	HN	THX	ABDM	PELV
AGD ↓				
Base	3.30%	2.70%	**1.80%**	1.30%
MI-Z	3.70%	3.00%	2.30%	1.70%
EI-Z	3.70%	2.70%	2.00%	1.30%
Crop-Z	3.70%	2.80%	2.50%	1.20%
MI	**3.00%**	2.60%	2.40%	1.40%
MI-TS	**3.00%**	**2.30%**	2.40%	**1.10%**
MGD ↓				
Base	**3.00%**	5.30%	3.80%	1.90%
MI-Z	7.00%	5.50%	**0.50%**	2.10%
EI-Z	6.20%	5.50%	0.60%	2.30%
Crop-Z	6.90%	5.40%	1.00%	2.30%
MI	3.80%	5.70%	0.60%	2.00%
MI-TS	**3.20%**	**5.20%**	3.20%	**1.50%**
QD ↓				
Base	12.90%	10.60%	**9.90%**	5.90%
MI-Z	13.40%	12.50%	11.80%	**5.90%**
EI-Z	**12.10%**	12.40%	11.70%	**5.90%**
Crop-Z	12.90%	12.30%	10.60%	**5.90%**
MI	12.50%	11.90%	10.90%	6.50%
MI-TS	**12.10%**	**10.20%**	10.20%	7.00%

5   Conclusions and Future Work

While the base model serves as a solid baseline, AP models like MI-Z, and MI-TS tend to reduce gender disparity in segmentation performance, especially in the ABDM region, indicating that these models are capable of handling the prior information to improve and mitigate gender-related anatomical differences. This study is limited by our small patient number and future work should be focused on testing different inputting prior strategies and their bias in larger datasets and architectures. Choosing the appropriate structures to add contextual information and mitigate should include the aid of clinicians. Studying other commonly recurring biases such as racial and size of the patient bias should be thought about as well.

References

1. Baskar, R., Lee, K.A., Yeo, R.M., Yeoh, K.W.: Cancer and radiation therapy: current advances and future directions. Int. J. Med. Sci. **9**, 193–199 (2012). https://api.semanticscholar.org/CorpusID:16909912
2. Puyol-Antón, E., et al.: Fairness in cardiac MR image analysis: an investigation of bias due to data imbalance in deep learning based segmentation. In: de Bruijne, M., et al. (eds.) MICCAI 2021. LNCS, vol. 12903, pp. 413–423. Springer, Cham (2021). https://doi.org/10.1007/978-3-030-87199-4_39
3. Brioso, R.C., Pedrosa, J., Mendonça, A.M., Campilho, A.: Semi-supervised multi-structure segmentation in chest X-ray imaging, pp. 814–820. IEEE (2023). https://doi.org/10.1109/CBMS58004.2023.00325
4. Buolamwini, J., Gebru, T.: Gender shades: intersectional accuracy disparities in commercial gender classification (2018). https://proceedings.mlr.press/v81/buolamwini18a.html
5. Chen, L.C., Zhu, Y., Papandreou, G., Schroff, F., Adam, H.: Encoder-decoder with atrous separable convolution for semantic image segmentation (2018). https://doi.org/10.48550/ARXIV.1802.02611. https://arxiv.org/abs/1802.02611
6. Huynh, E., et al.: Artificial intelligence in radiation oncology. Nat. Rev. Clin. Oncol. **17**(12), 771–781 (2020)
7. Isensee, F., et al.: Abstract: nnU-Net: self-adapting framework for U-net-based medical image segmentation. In: Bildverarbeitung für die Medizin 2019. I, pp. 22–22. Springer, Wiesbaden (2019). https://doi.org/10.1007/978-3-658-25326-4_7
8. Lee, T., Puyol-Antón, E., Ruijsink, B., Aitcheson, K., Shi, M., King, A.P.: An investigation into the impact of deep learning model choice on sex and race bias in cardiac MR segmentation. In: Wesarg, S., et al. (eds.) CLIP EPIMI FAIMI 2023. LNCS, vol. 14242. Springer, Cham (2023). https://doi.org/10.1007/978-3-031-45249-9_21
9. Lian, S., Li, L., Luo, Z., Zhong, Z., Wang, B., Li, S.: Learning multi-organ segmentation via partial- and mutual-prior from single-organ datasets. Biomed. Sig. Process. Control **80**, 104339 (2023). https://doi.org/10.1016/J.BSPC.2022.104339
10. Liu, Y., et al.: Head and neck multi-organ auto-segmentation on CT images aided by synthetic MRI. Med. Phys. **47**, 4294–4302 (2020). https://doi.org/10.1002/mp.14378
11. Liu, Z., et al.: Development and validation of a deep learning algorithm for auto-delineation of clinical target volume and organs at risk in cervical cancer radiotherapy. Radiother. Oncol. **153**, 172–179 (2020). https://doi.org/10.1016/J.RADONC.2020.09.060
12. Ma, C.Y., et al.: Deep learning-based auto-segmentation of clinical target volumes for radiotherapy treatment of cervical cancer. J. Appl. Clin. Med. Phys. **23** (2022). https://doi.org/10.1002/acm2.13470
13. Mancosu, P., et al.: Development of an immobilization device for total marrow irradiation. Pract. Radiat. Oncol. **11**(1), e98–e105 (2021)

14. Ronneberger, O., Fischer, P., Brox, T.: U-Net: convolutional networks for biomedical image segmentation (2015). https://arxiv.org/abs/1505.04597

15. Shen, C., et al.: Anatomical attention can help to segment the dilated pancreatic duct in abdominal CT. Int. J. Comput. Assist. Radiol. Surg. **19**, 655–664 (2024). https://doi.org/10.1007/s11548-023-03049-z

16. Song, Y., et al.: Automatic delineation of the clinical target volume and organs at risk by deep learning for rectal cancer postoperative radiotherapy. Radiotherapy Oncol. **145**, 186–192 (2020). https://doi.org/10.1016/J.RADONC.2020.01.020

17. Tong, N., Gou, S., Niu, T., Yang, S., Sheng, K.: Self-paced DenseNet with boundary constraint for automated multi-organ segmentation on abdominal CT images. Phys. Med. Biol. **65** (2020). https://doi.org/10.1088/1361-6560/ab9b57

18. Wasserthal, J., Meyer, M., Breit, H.C., Cyriac, J., Yang, S., Segeroth, M.: TotalSegmentator: robust segmentation of 104 anatomical structures in CT images (2022)

19. Wong, J.Y.C., et al.: Total marrow and total lymphoid irradiation in bone marrow transplantation for acute leukaemia. Rev. Lancet Oncol. **21**, 477–87 (2020)

20. Yang, J., Veeraraghavan, H., van Elmpt, W., Dekker, A., Gooding, M., Sharp, G.: CT images with expert manual contours of thoracic cancer for benchmarking auto-segmentation accuracy. Med. Phys. **47** (2020). https://doi.org/10.1002/mp.14107

EPIMI

Assessing the Impact of Sociotechnical Harms in AI-Based Medical Image Analysis

Emma A. M. Stanley[1,2,3,4]($\boxtimes$)(iD), Raissa Souza[1,2,3,4](iD), Anthony J. Winder[2](iD), Matthias Wilms[2,3,4,5](iD), G. Bruce Pike[2,3](iD), Gabrielle Dagasso[1,2,3,4](iD), Christopher Nielsen[1,2,3,4], Sarah J. MacEachern[3,4,5], and Nils D. Forkert[2,3,4](iD)

[1] Biomedical Engineering Graduate Program, University of Calgary, Calgary, Canada
emma.stanley@ucalgary.ca
[2] Department of Radiology, University of Calgary, Calgary, Canada
[3] Hotchkiss Brain Institute, University of Calgary, Calgary, Canada
[4] Alberta Children's Hospital Research Institute,
University of Calgary, Calgary, Canada
[5] Department of Pediatrics, University of Calgary, Calgary, Canada

Abstract. Clinical decision-making and radiology will inevitably be transformed by artificial intelligence (AI) in the coming years. The rapid adoption of AI in this domain, and in other everyday applications, has brought an increased awareness of the potential impacts and negative consequences that may occur throughout the sociotechnical systems that these technologies are implemented in. In this paper, we review and apply a previously published taxonomy of the sociotechnical harms of AI to investigate how these harms could manifest during the development and clinical implementation of AI-based medical image analysis. Through an illustrative case study example on computer-aided diagnosis using brain magnetic resonance imaging, we demonstrate how performing impact assessments of sociotechnical harms can assist in operationalizing the medical ethics principle of non-maleficence, thereby guiding the ethical development and implementation of AI technologies in healthcare.

1 Introduction

Despite great promise of the benefits of AI-based medical image analysis (MedIA) systems, those involved with research, development, implementation, and regulation of this technology still have ethical concerns to grapple with that hinder its responsible use in clinical practice [6]. For example, concerns about the fairness of MedIA models have become increasingly important to consider, since these systems carry the risk of propagating and amplifying societal inequalities. As the relationship between humans and advanced AI is still in its infancy, we are just starting to identify the potential sociotechnical impacts of algorithmic systems and we have yet to fully understand their consequences.

Based on a scoping review of research on modern algorithmic systems, Shelby et al. [28] published a general taxonomy of sociotechnical harms, which was an

E. Puyol-Antn et al. (Eds.): FAIMI 2024/EPIMI 2024, LNCS 15198, pp. 163–175, 2025.
https://doi.org/10.1007/978-3-031-72787-0_16

important step towards clearly defining the possible negative impacts of AI in various use cases. This taxonomy includes representational, allocative, quality of service, interpersonal, and social system harms. Their structured framework provides a comprehensive view of the ways in which algorithmic systems can cause harm, particularly through the entrenchment of societal inequities. A key motivation behind their work was to define potentially negative impacts that arise from the unique interactions between society and algorithmic technology so that AI developers and users are better aware of these harms, and thus can account for them in the development and implementation of these systems.

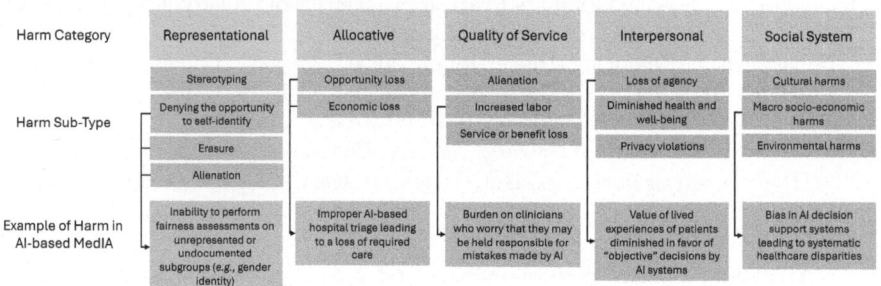

Fig. 1. Sociotechnical harm taxonomy and examples from AI-based medical image analysis (MedIA). Figure adapted from Shelby et al. [28].

Shelby et al. [28] discuss their taxonomy of sociotechnical harms in the context of various common algorithmic systems, including references to social media algorithms, speech recognition systems, image captioning, and financial modeling. Due to the broad scope of their review, and perhaps the relative dearth of published literature compared to other topics, AI for healthcare is not comprehensively discussed, let alone AI systems for radiology. Unlike social media algorithms and chatbots that interact directly with users, AI for MedIA stands out from other AI-based systems because a human intermediary (*i.e.,* a clinician) is a fundamental part of its implementation for clinical decision-making. This creates a distinct relationship between society and this application of AI, resulting in unique risks for sociotechnical harms compared to those from direct user-facing AI.

By considering the types of harms that could arise from the use of a particular algorithmic system in the context it is designed for, researchers and developers of AI can consider preventative safeguards, both in how the system is designed and how it should be operated. In the medical context, the practice of performing impact assessments to identify and mitigate the sociotechnical harms of AI for MedIA could be viewed as a way to operationalize the medical ethics principle of non-maleficence, which is better known by the phrase "do no harm". In this paper, we briefly review and discuss the taxonomy of sociotechnical harms by Shelby et al. with a special focus on AI-based MedIA systems, both in research and development (particularly in the context of algorithmic fairness assessments) and in clinical use (Fig. 1). We then present a hypothetical case study for computer-aided diagnosis of Parkinson's disease using multimodal brain MRI and describe the results of an impact assessment of sociotechnical harms for this system, showing how this process can help to identify avenues for operationalizing non-maleficence in the clinical implementation of AI.

2 Sociotechnical Harms in the Development and Use of Medical Imaging AI

2.1 Representational Harms

Representational harms are defined as the reinforcement of unjust societal hierarchies through the use of algorithmic systems [3,28]. These include *stereotyping*, which refers to how an algorithmic system's outputs reflect pre-defined "beliefs about the characteristics, attributes, and behaviours of members of certain groups", and *reifying essentialist social categories*, which means reinforcing the perception that socially constructed classifications of people are inherent, unchanging, and natural [28]. In the development of MedIA systems, including the study of how algorithmically "fair" these models are, researchers are mostly limited to analyzing the sociodemographic categories available for the data. These categories of subgroups, typically defined during data collection, are often broad (*e.g.*, "White", "Black", "Asian", "Other"), and assume that individuals can be defined as belonging exclusively to one group. After performing fairness assessments, statements such as "this model performs 10% worse on subgroup X" are often made. However, conclusions such as these imply homogeneity among members within these broad categories, which can often be socially constructed sensitive attributes such as race or socioeconomic status. This further lends itself to the implication that such performance disparities are due to some intrinsic property that makes someone belong to a particular subgroup, and not to potential adverse effects that arise from being racialized, for example [16]. Since fairness is typically analyzed with respect to demographic groups, the ultimate interpretation is that the model is biased against particular social groups [9,26], even though performance disparities could very likely be attributed to proxy variables (*e.g.*, historical healthcare expenditure [21], medical imaging scanner model [31], *etc.*). Thus, although algorithmic fairness assessments are highly important for evaluating safety and usability of MedIA AI systems, they nevertheless may result in a form of algorithmic stereotyping and/or reification of social groups. *Erasure, denial of the opportunity to self-identify,* and *alienation* are also distinct dimensions of representational harms, but are highly interconnected in the context of AI for MedIA. Shelby et al. [28] define erasure as the absence or systematic under-representation of certain social groups, to the extent that such groups are not even "legible" to algorithmic systems. In MedIA, this can include certain populations that are so under-represented in medical imaging datasets used for training (*e.g.*, Indigenous peoples), that they are simply grouped into the category of "Other" when it comes to assessing algorithmic fairness. Economically marginalized populations, including those in low and middle-income countries, as well as unhoused or under-resourced people in high-income countries are also more likely to be absent or only marginally included in datasets used to develop MedIA models, [1, 30] and thus less likely to benefit from such systems. Denying the opportunity to self-identify can also lead to erasure of social groups that should be considered in fairness assessments- for instance, in the context of gender identity. Most medical image datasets used to develop MedIA systems may include a sex variable but typically fail to report gender information in their metadata. Thus, the fairness of AI developed with these datasets can only be analyzed in the context of sex, even though it is known that gender is a social determinant of health [19] which could also be associated with confounders and algorithmic bias. Even within the category of sex, most large-scale medical imaging datasets used for research contain binary values of male or female, without considering intersex individuals, for example. The inability to self-report aspects of identity which are important to patients can also lead to alienation - the feeling that one "does not belong" due to being part of a certain social group [28]. This could also occur if an individual does not identify with any of the pre-defined demographic categories required when providing personal information (either when contributing data to a repository, or if that information is required as additional input to a model used in practice) and if, for example, forced to select an answer such as "other".

2.2   Allocative Harms

Allocative harms encompass those that result from uneven distribution of algorithmic decisions between populations [3, 28]. In the context of MedIA AI, these harms are most relevant when models are deployed for use in clinical practice. For instance, *opportunity loss* can occur when algorithmic systems facilitate different levels of access to resources, such as misdiagnosis with relevance for treatment [11] or improper triage [32] by a clinical decision support tool, which leads to a loss of necessary health care. *Economic loss* can arise from the financial burdens associated with opportunity loss, such as needless expenses and time lost due to a false positive diagnosis or increased long term financial burden due to a false negative diagnosis. Notably, AI-based computer aided-diagnostic systems for detecting breast cancer from mammograms have been shown to be prone to high false positive rates [8]. Both opportunity and economic loss often manifest in existing societal patterns of inequality, making it relevant in the well-known issues of MedIA systems that have been shown to produce, for example, systematically higher underdiagnosis rates for females, racial minorities, and those of lower socioeconomic status [26]. However, allocative harms could also occur independent of social groups, and instead as a result of technical artifacts or infrastructure. For instance, MedIA systems trained to detect pneumothorax from chest x-rays have been found to use thoracic tubes visible on the images as a shortcut [24], meaning that the system is likely to have poor sensitivity for any patients without a drain visible on their imaging. While this uneven distribution of algorithmic performance may not necessarily manifest along existing lines of inequity, it may still lead to disproportionate negative consequences for a subset of the patient population.

2.3 Quality of Service Harms

Quality of service harms are related to allocative harms, but specifically consider harms that occur as a result of systematic performance disparities between groups of people within society. A form of *alienation*, distinct from its form as a representational harm, occurs within this category as a "specific self-estrangement experience at time of technology use" [28]. For example, in a scenario where a clinician uses an AI-based MedIA tool, a patient may experience alienation if made aware that the specific model does not perform well for a particular demographic category they fall into. *Increased labor* is another dimension of quality of service harms, which Shelby et al. [28] describe as a greater burden on members of a certain social group to make an algorithmic system work as well for them as for others. In the context of AI-based MedIA, this increased labor may arise for the clinician using the AI system rather than the patient receiving care. For instance, if a clinician is informed that a system does not work as well for a particular demographic group, they may put increased effort into accounting for the shortcomings of that system when treating a patient with those demographic characteristics. Finally, *service or benefit loss* is the "loss of benefits of using algorithmic systems with inequitable system performance based on identity" [28]. In other words, the benefits of more accurate and efficient radiological workflows that AI-based MedIA has to offer may be offset or overshadowed by impacts on both clinician labor and systematic performance disparities affecting patient populations. Like allocative harms, quality of service harms could also occur regardless of whether algorithmic bias exists in MedIA systems. For example, any patient could feel alienated if their clinician chooses only to study and interact with the MedIA system on their computer screen, rather than bring them into the dialogue about their diagnosis or treatment. Furthermore, even if a MedIA system does not have known performance disparities between demographic groups, clinicians will still be subject to increased labor and mental load if they worry about the AI making mistakes, question their own judgement if it disagrees with the system, or fear being liable for the consequences of errors that the system makes [12].

2.4 Interpersonal Harms

Interpersonal harms describe situations where algorithmic systems negatively influence relationships between individuals or communities. Within this category, *loss of agency* refers to reduced autonomy due to how algorithmic systems are used. Diminished autonomy is especially at risk when AI systems are viewed as an objective, omniscient expert when aiding in healthcare decisions, with knowledge equal or superior to that of clinicians [15]. This "algorithmic paternalism" [17] can act as a barrier that prevents patients from questioning suggestions made by MedIA systems, especially when clinicians (who already have existing expert authority over a patient) agree with suggestions made by the system. Similarly, the value of lived experiences as valuable information when considering diagnoses or treatments may be diminished in favor of suggestions made by an "all-knowing" AI with "objective" evidence [17]. This loss of agency can impact the patient population as a whole, but may be particularly harmful to social groups who already experience dismissal of health concerns due to (implicit or overt) racism, sexism, ableism, or other discrimination [7]. *Diminished health and well-being* is another type of interpersonal harm that negatively impacts the well-being of social groups relative to other groups and institutions they interact with. Systematic underperformance of MedIA systems on certain sociodemographic subpopulations is an apparent example, which could lead to disproportionate levels of poorer health, financial burden, and reduced quality of life. Diminished health and well-being can also be associated with the psychological impacts of representational harms [28], which are likely to disproportionately impact marginalized groups. *Privacy violations* encompass a myriad of harms, including active surveillance, the feeling of being surveilled, and collection of data without informed consent. These have potential to occur at all points throughout the life cycle of an AI-based MedIA system. For example, in Canada, explicit consent is not required to use medical imaging data collected as part of a clinical exam for secondary purposes such as AI system development, as long as the data is de-identified [23], even though some studies have shown that it is possible to re-identify such images [22]. In the process of training MedIA systems, especially with distributed learning approaches, which perform training across multiple machines with smaller sample sizes to avoid explicit data-sharing [29], there is a risk of adversarial attacks that can result in theft of medical image data [20] or the identification of sensitive characteristics about individuals [33]. A further type of privacy violation may simply be that patients are uncomfortable with their personal medical images being accessed by AI or used widely to train MedIA systems, which may go on to be commercialized [18]. This discomfort may be linked to a lack of trust in the specific technology as well as in corporate tech oligopolies that may have control over such systems and profit from the use of patient data.

2.5   Social System Harms

Societal harms are those that reflect the "macro-level effects" of algorithmic systems that perpetuate or amplify bias, inequality, and disproportionate power dynamics at a large scale [28]. *Cultural harms* introduced by AI can impact cultural safety, stability, and values, which could be considered within the culture of medical practice. Many clinicians may consider patient autonomy and shared decision-making as key values in their profession [27], but the use of AI systems in clinical practice may result in a loss of dialogue between patients and their healthcare providers [15]. Radiologists could also become over-reliant on MedIA systems and favour AI-generated advice over their own judgement [5]. This automation bias could change the way radiologists perform their work and potentially lead them to question the value of their training and profession. The widespread implementation and paternalistic framing of AI-based MedIA could also further dismiss value in non-Western and/or Indigenous medical practices, which have been historically stigmatized and even criminalized [14]. *Macro socio-economic harms* that "exacerbate digital divides" and "entrench systemic inequalities" [28] are particularly relevant when considering who develops MedIA AI, and who it is developed for. With a large part of research and development originating in high-income countries, using data from privileged populations [30], the implementation of MedIA systems is likely to benefit those populations the most [2]. However, even well-meaning initiatives to collect medical imaging data from under-resourced settings could result in forced collection of data and exploitation of those in low- and middle-income countries ("data colonialism") [1]. Furthermore, there is a risk of a feedback loop of inequity occurring, where those who get access to the benefits of AI get healthier, and those without access end up with worsened health outcomes. For instance, many AI-based MedIA models depend on measuring biomarkers (*e.g.,* with MRI), which are too expensive to be the widespread standard of care worldwide. Thus, in nations where such imaging procedures are not publicly funded or available, wealthy individuals who can afford them may receive better and more accurate medical care compared to those who cannot. The introduction or exacerbation of healthcare and economic disparities at a large scale due to algorithmic bias in MedIA systems is also a significant macro socio-economic harm. Finally, the computational resources required to develop and apply models on large-scale, high dimensional medical imaging data, multiplied by the hundreds of academic and industry labs in the domain contribute to *environmental harm* through carbon emissions, mineral extraction, and cooling water required to maintain compute facilities [25].

3   Case Study: Sociotechnical Harm Impact Assessment in a Parkinson's Disease Diagnostic Model

3.1   Methods

We performed a collaborative impact assessment of sociotechnical harms for an illustrative example of an AI-based MedIA system that is to be implemented in clinical practice. The system in this case example is based on the Parkinson's disease classification model published by Camacho et al. [4]. For the impact assessment, we *i)* identified affected groups, *ii)* identified potential harms, considering the taxonomy of Shelby et al. [28], and *iii)* identified potential actions that could be taken to prevent or mitigate such harms. The group involved in the impact assessment was comprised of eight researchers (four doctoral students, one research assistant, three faculty members) with expertise on medical imaging and/or development of AI-based MedIA.

3.2   Case Example Background

A convolutional neural network-based deep learning model is to be used for computer-assisted diagnosis of Parkinson's disease (PD). The model inputs are structural T1-weighted MRI and diffusion tensor imaging (DTI), as well as age and sex information. It was trained and validated on images from 1,264 individuals originating from eight separate studies across Canada, the United States, the United Kingdom, and Germany, and had a rather balanced representation of diagnosed PD (n = 611) compared to healthy (n = 653) participants. Both healthy and PD groups had a majority of male participants, with an overall female representation of 37.5%. The average age was 65.5 years for the PD group and 66.1 years for the healthy group. Train, validation, and test splits were 75%, 10% and 15%, stratified by age, sex, PD/healthy status, and data study. All datasets were acquired on 3T scanners, from three different manufacturers. The model achieved 80.8% accuracy, 82.4% specificity, and 79.1% sensitivity on the test data. Although Camacho et al. [4] did not include fairness assessments in their paper, for this case study, we assume that a sex subgroup analysis was performed, which resulted in a sensitivity of 84% for males and 71% for females, and specificity of 80% for males and 86% for females. Furthermore, we assume that this system is meant to be used as an additional tool to assist in early diagnosis of PD in a movement disorder clinic in Calgary, Canada.

3.3   Impact Assessment Results

We identified a wide variety of affected groups, including patients, patient families, radiologists, neurologists, researchers and developers of the system, patient advocacy groups, funding organizations, clinic administration, and ethics boards. However, the discussion primarily focused on concerns relevant to patients, their families, neurologists, and radiologists (i.e., those directly involved or affected by the use of this system).

After considering affected groups, we began discussing concerns associated with the development and use of the system, the potential downstream impact of these concerns, and then considered which groups those concerns were relevant to as well as which sociotechnical harm subtype they could be categorized into. These results are summarized in Fig. 2. Our group's main concern was that sampling bias in the dataset and the limited scope of subgroup assessment could cause performance disparities, leading to misdiagnoses in different patient groups. In addition, several concerns that were specific to this case example arose, including 1) uncertainty in how the system would perform on patients with prodromal or early stage PD, potentially leading to poor performance in those cases, 2) long wait times (≈5 months) to receive MRI scans in Calgary, which could prolong the diagnostic process for patients, and 3) the current standard for diagnosing PD relies on clinician experience, not structural MRI biomarkers, so introducing this AI-based diagnostic tool would significantly disrupt neurologists' current workflow.

Concerns	Impacts	Affected Groups	Relevant Harm Subtypes
Sampling bias in dataset (demographic composition, different image acquisition protocols/systems)	Risk of misdiagnosis due to algorithmic bias	Patients	Opportunity loss Economic loss
		Patient families	Diminished health and well-being
		Radiologists Neurologists	Increased labor
Model requires sex as an input variable	Patients who do not identify as male or female may feel uncomfortable/ unrepresented	Patients	Alienating social groups Erasing social groups Denying the opportunity to self-identify Reifying essentialist categories
Uncertainty in how the system performs across other sociodemographic categories	Neurologists may not recommend use of the system on certain patients if they don't know how well it performs	Patients	Service/benefit loss Opportunity loss Diminished health and well-being
		Radiologists Neurologists	Increased labor
Uncertainty in how the system performs on early stage/prodromal PD cases	Early stage/prodromal patients may have a higher risk of misdiagnosis	Patients	Opportunity loss Economic loss
		Patient families	Diminished health and well-being
		Radiologists Neurologists	Increased labor
Clinicians are used to a certain workflow for the standard of care, which gets disrupted by the integration of AI	Automation bias, underutilization, and/or frustration	Radiologists Neurologists	Service/benefit loss Increased labor Cultural harms
Costs associated with using the system, e.g., transportation to clinic and imaging centers (especially from rural areas)	Some individuals may have financial barriers to accessing diagnostic care, or experience financial hardship as a result	Patients	Opportunity loss Economic loss Macro socio-economic harms Diminished health and well-being
		Patient families	Economic loss Macro socio-economic harms
Requirement to get MR imaging to receive a diagnosis	Patients must wait for extended time periods to receive a final diagnosis	Patients	Service/benefit loss Opportunity loss
		Patient families	Diminished health and well-being
Use of patient imaging for continued evaluation of the system, even after potential death of the patient	Discomfort in medical images being used to train/evaluate AI systems	Patients	Privacy violations
		Patient families	

Fig. 2. Summary of concerns and corresponding harms identified in the collaborative impact assessment on AI-assisted diagnosis of Parkinson's disease from brain MRI. Please note that this is not intended to be an exhaustive list of all relevant harms associated with this case example.

Rather than discussing specific mitigation strategies for each harm, we identified that two broad approaches that could prevent numerous harms. First, "silent trials" [10] would enable active monitoring of the system without affecting clinical decisions. With additional collection of demographic data during this phase, the team of clinicians and developers who are assessing this system can check for previously unseen demographic and image acquisition biases. If such biases exist, further modifications such as fine tuning, data harmonization, or bias mitigation could be employed. This would also provide the opportunity to get feedback from clinicians regarding how the system fits into their workflow, helping to prevent harms associated with automation bias or underutilization. Second, public outreach and education about the system and its use could help to inform patients and their families about how exactly their data is used by AI, potentially dispelling privacy concerns. This approach could also help to identify whether there are patient groups that could use additional social assistance in receiving this care (*e.g.,* free transportation to the clinic).

4 Conclusion

In this paper, we briefly discussed potential sociotechnical harms that may occur as a result of the development and deployment of AI-based MedIA systems. Harms that are typically discussed in the context of these systems are often limited to data privacy and group unfairness resulting from algorithmic bias. However, by analyzing the taxonomy proposed by Shelby et al. [28], we identified numerous other avenues through which harms could occur within the broader sociotechnical scope of AI-based MedIA. As with previous literature on sociotechnical harms [3,28], many of the considerations we discussed align with existing patterns of societal inequity - especially within representational harms, and harms resulting from systems that produce performance disparities between subpopulations. Nevertheless, we also note that several sociotechnical harms that we identified can occur irrespective of existing societal marginalization - for instance, all clinicians using MedIA systems for decision support and any patients receiving AI-assisted care may experience harms as a result of the integration of these systems in clinical practice. Using knowledge of the sociotechnical harm taxonomy, we performed a collaborative impact assessment for identifying ways to minimize the risk of harm from the hypothetical use of an AI-based Parkinson's disease early detection system. This exercise encouraged each participant to consider sociotechnical impacts beyond the scope of just the AI system, and enabled the discovery of unique manifestations of sociotechnical harms that could arise in the case example we considered. Those who participated in the impact assessment come from various academic and cultural backgrounds. However, other perspectives (*e.g.,* lawyers, ethicists) may have readily identified more concerns and potential sociotechnical harms related to other affected groups. We are looking forward to integrating these other areas of expertise in future discussions to develop a more comprehensive impact assessment strategy for identifying a broader range of harms and mitigation strategies. Additionally, some subtypes in each harm category from the original taxonomy were not discussed in this paper due to not being identified as relevant in the context of AI-based MedIA, but our failure to identify these does not imply that they do not exist. Finally, an impact assessment of sociotechnical harms should be considered as only a small piece of a comprehensive framework (see [13,16]) for the development and implementation of responsible and ethical AI for medical image analysis. While the fault for any number of the sociotechnical harms discussed in this work cannot be placed squarely on a single person or institution, those who research, develop, market, and use algorithmic MedIA systems should, at least, be aware of them - and aware that they can implement measures to help ensure that these systems "do no harm".

Disclosure of Interests. The authors have no competing interests to declare that are relevant to the content of this article.

References

1. Ethics and governance of artificial intelligence for health: WHO Guidance (2021)
2. Arora, A., et al.: The value of standards for health datasets in artificial intelligence-based applications. Nat. Med. **29**(11), 2929–2938 (2023)
3. Barocas, S., Crawford, K., Shapiro, A., Wallach, H.: The problem with bias: allocative versus representational harms in machine learning' (2017)
4. Camacho, M., Wilms, M., Almgren, H., Amador, K., Camicioli, R., et al.: Exploiting macro- and micro-structural brain changes for improved Parkinson's disease classification from MRI data. npj Parkinsons Dis. **10**(1), 1–12 (2024)
5. Dratsch, T., et al.: Automation bias in mammography: the impact of artificial intelligence BI-RADS suggestions on reader performance. Radiology **307**(4), e222176 (2023)
6. Geis, J.R., et al.: Ethics of artificial intelligence in radiology: summary of the joint European and North American multisociety statement. Insights Imaging **10** (2019)
7. Hildenbrand, G.M., Perrault, E.K., Rnoh, R.H.: Patients' perceptions of health care providers' dismissive communication. Health Promot. Pract. **23**(5), 777–784 (2022)
8. Houssami, N., Given-Wilson, R., Ciatto, S.: Early detection of breast cancer: overview of the evidence on computer-aided detection in mammography screening. J. Med. Imaging Radiat. Oncol. **53**(2), 171–176 (2009)
9. Klingenberg, M., Stark, D., Eitel, F., Budding, C., Habes, M., et al.: Higher performance for women than men in MRI-based Alzheimer's disease detection. Alzheimer's Res. Ther. **15**(1), 84 (2023)
10. Kwong, J.C.C., et al.: The silent trial - the bridge between bench-to-bedside clinical AI applications. Frontiers Digit. Health **4** (2022)
11. Lashbrook, A.: AI-driven dermatology could leave dark-skinned patients behind, August 2018
12. Lawton, T., et al.: Clinicians risk becoming 'liability sinks' for artificial intelligence. Future Healthc. J. **11**(1), 100007 (2024)
13. Lekadir, K., Osuala, R., Gallin, C., Lazrak, N., Kushibar, K., et al.: FUTURE-AI: guiding principles and consensus recommendations for trustworthy artificial intelligence in medical imaging. arXiv:2109.09658 [cs] (2021)
14. Li, R.: Indigenous identity and traditional medicine: pharmacy at the crossroads. Can. Pharm. J. (Ott) **150**(5), 279–281 (2017)
15. McCradden, M., Hui, K., Buchman, D.Z.: Evidence, ethics and the promise of artificial intelligence in psychiatry. J. Med. Ethics (2022). 2022-108447

16. Mccradden, M., et al.: What's fair is ... fair? Presenting JustEFAB, an ethical framework for operationalizing medical ethics and social justice in the integration of clinical machine learning: JustEFAB. In: 2023 ACM Conference on Fairness, Accountability, and Transparency, Chicago, IL, USA, pp. 1505–1519 (2023)

17. McCradden, M.D., Kirsch, R.E.: Patient wisdom should be incorporated into health AI to avoid algorithmic paternalism. Nat. Med. **29**(4), 765–766 (2023)

18. McKay, F., Treanor, D., Hallowell, N.: Inalienable data: ethical imaginaries of de-identified health data ownership. SSM - Qual. Res. Health **4**, 100321 (2023)

19. Miani, C., Wandschneider, L., Niemann, J., Batram-Zantvoort, S., Razum, O.: Measurement of gender as a social determinant of health in epidemiology-a scoping review. PLoS ONE **16**(11), e0259223 (2021)

20. Nielsen, C., Tuladhar, A., Forkert, N.D.: Investigating the vulnerability of federated learning-based diabetic retinopathy grade classification to gradient inversion attacks. In: Antony, B., Fu, H., Lee, C.S., MacGillivray, T., Xu, Y., Zheng, Y. (eds.) OMIA 2022. LNCS, vol. 13576, pp. 183–192. Springer, Cham (2022). https://doi.org/10.1007/978-3-031-16525-2_19

21. Obermeyer, Z., Powers, B., Vogeli, C., Mullainathan, S.: Dissecting racial bias in an algorithm used to manage the health of populations. Science **366**(6464), 447–453 (2019)

22. Packhäuser, K., Gündel, S., Münster, N., Syben, C., Christlein, V., Maier, A.: Deep learning-based patient re-identification is able to exploit the biometric nature of medical chest X-ray data. Sci. Rep. **12**(1), 14851 (2022)

23. Parker, W., Jaremko, J.L., Cicero, M., Azar, M., El-Emam, K., et al.: Canadian association of radiologists white paper on de-identification of medical imaging: Part 1, general principles. Can. Assoc. Radiol. J. **72**(1), 13–24 (2021)

24. Rueckel, J., Trappmann, L., Schachtner, B., Wesp, P., Hoppe, B.F., et al.: Impact of confounding thoracic tubes and pleural dehiscence extent on artificial intelligence pneumothorax detection in chest radiographs. Invest. Radiol. **55**(12), 792–798 (2020)

25. Selvan, R., Bhagwat, N., Wolff Anthony, L.F., Kanding, B., Dam, E.B.: Carbon footprint of selecting and training deep learning models for medical image analysis. In: Wang, L., Dou, Q., Fletcher, P.T., Speidel, S., Li, S. (eds.) MICCAI 2022. LNCS, vol. 13435, pp. 506–516. Springer, Cham (2022). https://doi.org/10.1007/978-3-031-16443-9_49

26. Seyyed-Kalantari, L., Zhang, H., McDermott, M.B.A., Chen, I.Y., Ghassemi, M.: Underdiagnosis bias of artificial intelligence algorithms applied to chest radiographs in under-served patient populations. Nat. Med. **27**(12), 2176–2182 (2021)

27. Shanafelt, T.D., Schein, E., Minor, L.B., Trockel, M., Schein, P., Kirch, D.: Healing the professional culture of medicine. Mayo Clin. Proc. **94**(8), 1556–1566 (2019)

28. Shelby, R., Rismani, S., Henne, K., Moon, A., Rostamzadeh, N., et al.: Sociotechnical harms of algorithmic systems: scoping a taxonomy for harm reduction. In: Proceedings of the 2023 AAAI/ACM Conference on AI, Ethics, and Society, AIES 2023, New York, NY, USA, pp. 723–741 (2023)

29. Souza, R., Stanley, E.A.M., Camacho, M., Camicioli, R., et al.: A multi-center distributed learning approach for Parkinson's disease classification using the traveling model paradigm. Front. Artif. Intell. **7** (2024)

30. Souza, R., Stanley, E.A.M., Forkert, N.D.: On the relationship between open science in artificial intelligence for medical imaging and global health equity. In: Wesarg, S., et al. (eds.) CLIP EPIMI FAIMI 2023. LNCS, vol. 14242, pp. 289–300. Springer, Cham (2023). https://doi.org/10.1007/978-3-031-45249-9_28

31. Souza, R., Winder, A., Stanley, E.A., Vigneshwaran, V., Camacho, M., et al.: Identifying biases in a multicenter MRI database for Parkinson's disease classification: is the disease classifier a secret site classifier? IEEE J. Biomed. Health Inf., 1–8 (2024)
32. Weisberg, E.M., Chu, L.C., Fishman, E.K.: The first use of artificial intelligence (AI) in the ER: triage not diagnosis. Emerg. Radiol. **27**(4), 361–366 (2020)
33. Wu, M., et al.: Evaluation of inference attack models for deep learning on medical data (2020). http://arxiv.org/abs/2011.00177

Practical and Ethical Considerations for Generative AI in Medical Imaging

Debesh Jha[1(✉)], Ashish Rauniyar[2], Desta Haileselassie Hagos[3],
Vanshali Sharma[1], Nikhil Kumar Tomar[1], Zheyuan Zhang[1], Ilkin Isler[4],
Gorkem Durak[1], Michael Wallace[5], Cemal Yazici[6], Tyler Berzin[7],
Koushik Biswas[1], and Ulas Bagci[1]

[1] Machine and Hybrid Intelligence Laboratory, Northwestern University, Evanston,
USA
Debesh.jha@northwestern.edu
[2] Sustainable Communication Technologies (SCT), SINTEF Digital, Trondheim,
Norway
[3] Department of Electrical Engineering and Computer Science, Howard University,
Washington, D.C, USA
[4] Department of Computer Science, University of Central Florida, Orlando, USA
[5] Mayo Clinic Florida, Jacksonville, USA
[6] Department of Medicine, University of Chicago, Chicago, USA
[7] Beth Israel Deaconess Medical Center, Harvard Medical School, Boston, USA

Abstract. Generative Artificial Intelligence (AI) has the potential to
transform medicine. It is helpful to clinicians and radiologists for diagnosis, screening, treatment planning, interventions, and drug development. It benefits the clinical flow with real-time decision-support systems. While generative AI can potentially improve healthcare, it also
introduces new ethical issues that require careful analysis and mitigation
strategies. This work emphasizes the ethical aspects of generative AI in
medical imaging, aiming to ensure that advancements in this field align
with established ethical principles and societal values. We delve into the
ethical implications surrounding bias, fairness, patient privacy, consent,
transparency, explainability, intellectual property, and data ownership.
Furthermore, we discuss regulations governing the use of synthetic medical data. To promote equitable application of these powerful tools, we
also propose clear guidelines for promoting fairness, mitigating bias, and
ensuring diversity within generative AI models.

Keywords: Medical imaging · Generative AI · Synthetic data ·
Ethical consideration · Fairness · Privacy · Transparency · Bias
mitigation · Patient consent

1 Introduction

Generative Artificial Intelligence (AI) is rapidly transforming different sectors
in healthcare, in particular medical imaging. The popular types of generative

E. Puyol-Antón et al. (Eds.): FAIMI 2024/EPIMI 2024, LNCS 15198, pp. 176–187, 2025.
https://doi.org/10.1007/978-3-031-72787-0_17

AI models particularly useful for image generation are diffusion models [15], Transformer-based models (GPT-4 [26]), DALL-E 3[1], Generative Adversarial Networks (GANs) [10], and Variational Autoencoders (VAEs) [18]. These models can generate diverse, realistic-looking, high-quality synthetic images or videos, including rare medical scenarios that might be important in the medical domain for an automatic disease detection system. The synthetic data can be used to train and evaluate other AI algorithms. This advancement can significantly improve disease diagnosis. It sidesteps the need for manual medical annotation efforts, accelerating model deployment. It can help identify complex disease mechanisms, predict clinical outcomes, and prescribe tailored patient treatments.

The examples of recent generative AI models for image generation include stable diffusion [33], Apple Intelligence[2] and Midjourney [24]. For natural language processing tasks, GPT-4o[3], Claude 3 family (Anthropic)[4], Gemini[5], LaMDA [6], PaLM [7], Bloom [19], LLaMA [34] are some of the most popular approaches. For programming tasks, OpenAI codex [5] and Alphacode [20] are frequently used. Related to medical imaging, the famous Medical Open Network for Artificial Intelligence (MONAI) framework [3] helps in the creation of large synthetic datasets for training AI models prioritizing patient privacy. By preserving patient anonymity and maintaining confidentiality, MONAI enables the ethical use of sensitive medical data, thus promoting ethical data usage practices [3].

Limited real patient data due to privacy concerns remains a significant hurdle in developing robust medical AI models. Figure 1 shows the some of the ethical and practical challenges in integrating generative AI in medicine. The MONAI framework facilitates the creation of large synthetic datasets while preserving patient anonymity [3]. This promotes ethical data usage and allows researchers to train models on a broader range of data without compromising privacy. Additionally, pre-trained generative models like *RadImageGAN* [23] empower researchers to synthesize realistic medical images that closely resemble real data. These models offer flexibility to generate customized datasets tailored to specific applications. Advancements in generative models have a two-fold benefit: they significantly increase the availability of training data for medical AI models, and they democratize AI in biomedical imaging by providing access to data and expertise previously limited to specialized domains. This fosters wider adoption and innovation in the field [23].

Beyond data generation, generative models can enhance the quality of existing medical images by denoising and enhancing them using techniques like VAEs [35]. This leads to more precise visualization of subtle anatomical details for radiologists, potentially aiding in faster and more precise diagnoses [35]. Additionally, generative AI can be useful in automating lesion segmentation tasks.

---

[1] https://openai.com/index/dall-e-3/.
[2] https://www.apple.com/apple-intelligence/.
[3] https://openai.com/index/hello-gpt-4o/.
[4] https://claude.ai/new.
[5] https://gemini.google.com/app.

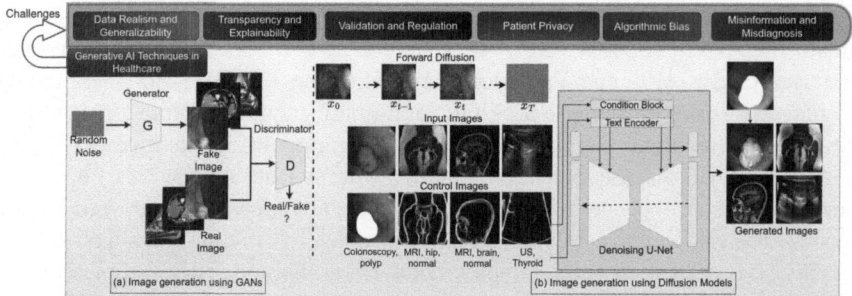

Fig. 1. Ethical and practical challenges in integrating generative AI (synthetic image generation technique) with medical imaging. Part (a) shows the GANs [10], and part (b) shows diffusion models [29] producing high-fidelity synthetic medical images.

Training AI models can achieve this to differentiate healthy and diseased tissues, significantly assisting radiologists in faster and more precise diagnoses [22]. Furthermore, the potential for personalized medicine is another opportunity for generative AI in medical imaging. The generative AI models create synthetic data tailored to particular diseases or responses to treatment. This data can then be leveraged to develop individualized treatment plans and predict patient outcomes with greater accuracy [21].

While these advancements in generative AI models open new opportunities to enhance healthcare outcomes through better data availability and improved model performance, they also raise significant ethical concerns. One critical concern is ensuring the authenticity and *representativeness* of synthetic data. If these images aren't realistic reflections of actual patients, they could mislead AI models, leading to biased diagnoses and potentially harmful treatment decisions [1]. To address this, researchers should develop methods to evaluate the authenticity and applicability of synthetic data. Additionally, mitigating biases in generative models through diverse real-world datasets and fairness-aware learning approaches is crucial to prevent biased diagnoses and treatment recommendations [1]. Another concern is data privacy and patient rights [36]. Even anonymized patient data used for synthetic image generation requires informed consent. Open and transparent communication regarding data usage policies is important [1]. Implementing robust anonymization techniques and establishing strong governance frameworks can further ensure patient privacy protection [36]. By addressing these ethical considerations, we can ensure the responsible development and deployment of generative AI models in medical imaging, ultimately leading to improved patient privacy and equitable healthcare systems.

The main contributions of this work are as follows:

- We highlight practical ethical concerns (patient privacy, algorithmic bias, misinformation, or misdiagnosis) related to generative AI models in medicine. Additionally, we provide potential guidelines to address these concerns, such as ethical generation and use of synthetic data, data realism, generalization, fairness, privacy, consent, transparency, explainability, safety, and intellectual property.

- Our study highlights some of the challenges related to privacy and security risks and performance drops associated with utilizing diffusion model-generated synthetic images in medical imaging. The example of specific vulnerabilities can be observed in Fig. 2 and Fig. 3).
- We examine the impact of adversarial perturbations on medical image segmentation performance. When adversarial perturbations were introduced, we observed a significant performance drop on the Liver MRI dataset (see Fig. 4).
- We discuss regulations on synthetic medical data and propose techniques to mitigate them. We also discuss measuring fairness, bias, and diversity, ensuring equitable outcomes for all patients. By prioritizing fairness, safety and interpretability, generative AI can have a positive societal impact.

2   Synthetic Image Generation

Synthetic image generation offers a promising avenue for addressing data scarcity in medical AI. These techniques produce artificial images that mimic real-world clinical data like X-rays, CT scans, and MRIs [13]. This synthetic data can be used for various purposes, including training AI models with limited real data, simulating rare or complex medical conditions, and facilitating diverse case studies for research and education [13].

However, ethical concerns surround the potential for bias in synthetic medical images. Generative AI algorithms can inadvertently inherit biases present in their training data, leading to the generation of unfair or biased synthetic images [14]. Addressing these biases is critical to ensure fairness and equity in healthcare applications, particularly for diagnosis and treatment decisions. Additionally, concerns arise when synthetic images fail to capture the full complexity and nuances of real medical data. This can lead to inaccurate conclusions or decisions in clinical practice. Therefore, it is essential to address these ethical implications to ensure the responsible development and ethical use of synthetic image generation in healthcare.

Diffusion models are emerging as powerful tools for synthetic medical image generation. However, recent work by Carlini et al. [4] suggests they may pose a greater privacy risk compared to GANs due to their tendency to memorize training data more extensively (approximately 2x). This raises concerns about potential data leakage, particularly when training data includes sensitive patient information like medical histories, personal identifiers, or genetic/biometric records.

Standard training procedures involve masking private details from training images before feeding them into the model. However, instances like those depicted in Fig. 2 raise questions. Here, synthetic images generated by diffusion models contain text alongside masked textual details. Do these texts represent:

- **Unmasked Data Leakage:** Was sensitive information overlooked during the anonymization process?

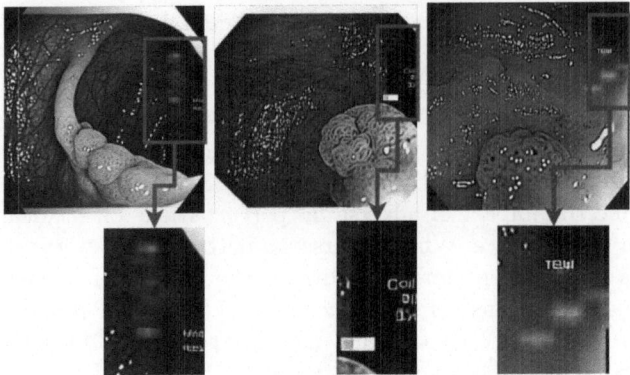

Fig. 2. Sample synthetic images generated using a diffusion model. The accompanying text within these images is not completely masked, which could be a concern if it leaks some sensitive information.

- **Model Decoding Attempts:** Is the model attempting to reconstruct the masked text?

These findings and unanswered questions highlight the need for further investigation into potential privacy vulnerabilities arising from training data or the image generation process itself.

Image generative models suggest pre-trained foundation models in medical AI workflows for enterprises. The ethical use of image-generative models requires ensuring that the generated images accurately represent real medical conditions. Poorly generated images could lead to incorrect diagnoses or treatment decisions, potentially harming patients [25]. Therefore, developers must prioritize the quality and accuracy of generated images to uphold patient safety and well-being. Although AI-generated synthetic images promise to revolutionize medical imaging through improved diagnostic accuracy and patient outcomes, their development involves significant challenges. These challenges include computational demands, data availability, training data quality, and model performance [38]. Generative AI models require billions of parameters and extensive training datasets, resulting in high computational costs. However, ensuring unbiased and high-quality data remains critical but challenging due to limited data availability in the medical domain [38]. Additionally, restricting access to existing public datasets further complicates the training process. Moreover, applying synthetic images in medical imaging also introduces the following practical and ethical challenges that need to be addressed [9].

2.1 Practical Challenges

Data Realism and Generalizability. Biases within real patient data used for training can be amplified in synthetic images, potentially leading to models that underperform on diverse patient populations. The ability of these models

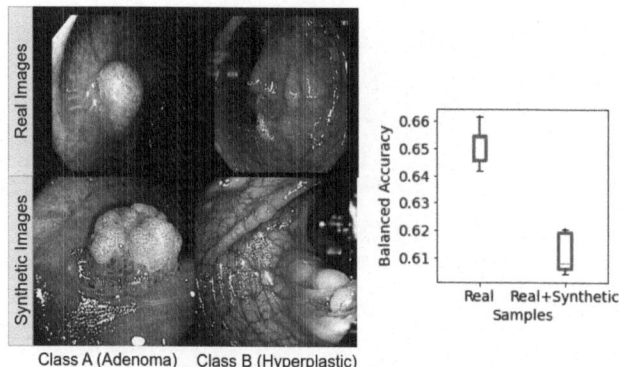

Fig. 3. The figure presents a) real samples and synthetic samples generated using a diffusion model, and b) a boxplot depicting the comparison of classification outcomes in two scenarios, one when only real images are used for training and the case when synthetic images were introduced in the training set. The results demonstrate how visually appealing synthetic images can reduce the performance of a classifier designed to identify pathological classes of polyps when combined with real ones.

to capture the full range of disease presentations also needs to be established to avoid missed diagnoses. As shown in Fig. 3, the visually appealing synthetic images fail to capture the pathological aspect of medical images, degrading the classification results.

Transparency and Explainability. Unlike traditional medical images, synthetic images lack inherent context. Understanding how the AI model generates a specific image is critical for healthcare professionals to assess its validity. A lack of transparency can limit trust and adoption among clinicians.

Validation and Regulation. Existing validation processes for medical imaging tools need to be adapted to encompass synthetic images. Regulatory frameworks are necessary to ensure the safety and efficacy of AI-generated synthetic images before widespread clinical use. These frameworks should address both technical and ethical concerns, including data privacy and biases.

2.2   Ethical Challenges

Patient privacy. Synthetic image generation relies on real patient data, raising concerns about privacy and potential misuse. De-identification techniques might not be sufficient to guarantee anonymity, and the possibility of re-identification through advanced AI techniques necessitates robust data governance practices for protecting sensitive medical information.

Algorithmic Bias. Biases inherent in the training dataset may persist in the synthetic images, resulting in misdiagnosis and unfair treatment, especially for underrepresented patients. It is particularly observed in cases where there are underrepresented patient groups. To mitigate bias, thorough data selection must

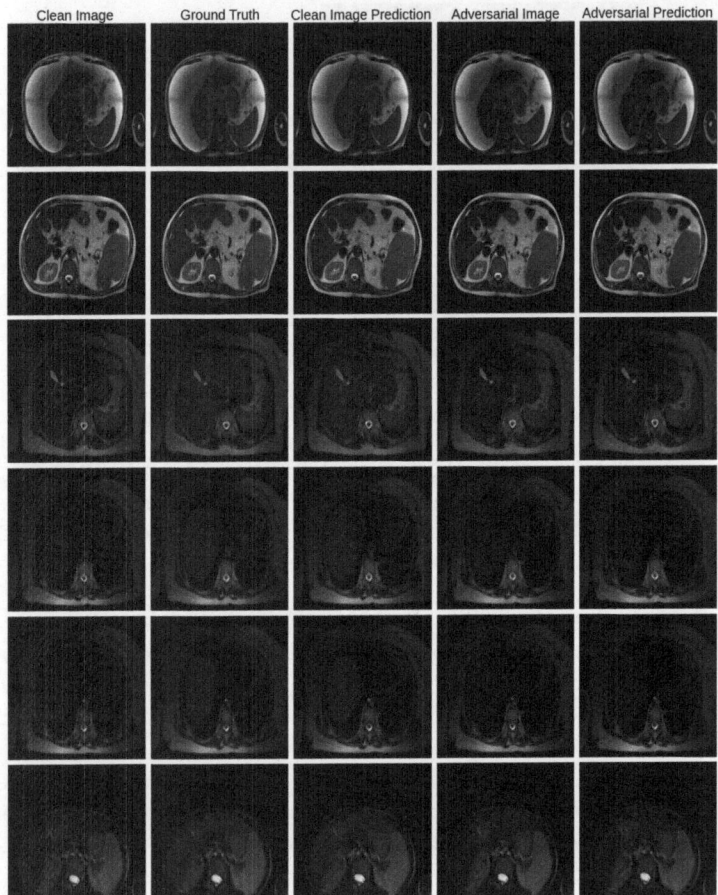

Fig. 4. The figure shows (a) Clean T2-weighted Liver MRI image, (b) Ground truth, (c) Clean image prediction by UNet [30] model, (d) Adversarial image generated by using Fast Gradient Sign Method (FGSM), and (e) Adversarial Prediction.

be done, diverse and representative data must be included and continuous monitoring of generative AI models are required.

Misinformation and Misdiagnosis. The realistic nature of synthetic images raises concerns about their potential to generate misleading medical information. Thus, clear communication strategies are crucial to ensure responsible use and accurate interpretation, particularly with a patient's medical history [1].

3   Effect of Adversarial Permutation

Fig. 4 shows the impact of adding adversarial perturbations on segmentation performance. We utilized the Fast Gradient Sign Method (FGSM) [11] to generate the adversarial examples and employed the UNet [30] model for T2-weighted

Liver MRI segmentation. The UNet achieved a dice coefficient of 88.49%, mIoU of 82.54%, recall of 89.96%, precision of 90.80%, and Haudorff distance (HD) of 3.5421. However, when adversarial input was introduced, the model's performance significantly dropped, with a dice coefficient of 35.13%, mIoU of 27.95%, recall of 31.49%, precision of 60.41%, and HD of 6.60.

The experiments were carried out using the PyTorch framework [27]. For the segmentation tasks, a batch size of 16 and a learning rate of 1e-4 were considered. The network was set to train up to 500 epochs of training with an early stopping set to 50 to fine-tune the parameters. To further enhance network performance, we utilized a hybrid loss function that combined binary cross-entropy and dice loss, along with an Adam [17] optimizer for parameter updates. The data was partitioned into three subsets: 80% for training, 10% for validation, and 10% for testing. To balance the training time and model complexity, we resized the image to 256 × 256 pixels in plane resolution. All experiments were conducted on the A100 GPU server.

The qualitative results demonstrate that both original and adversarial modified images appear identical to human observers and even radiologists. However, segmentation output on contrast enhanced abdominal MRI images (T2-weighted Liver MRI) was affected significantly. This highlights the need to develop robust deep learning strong models and defense techniques to withstand minor but powerful pixel-level alternations. Some techniques to enhance the model's robustness are strong data augmentation, adversarial training, smooth activation function, advanced regularization, and distillations.

4   Regulations on Synthetic Medical Data

Synthetic data addresses the current challenges of medical data availability. Training AI models using such datasets also complies with medical privacy regulations such as the Health Insurance Portability and Accountability Act (HIPAA) and General Data Protection Regulation (GDPR). *Ensuring compliance, privacy, and data protection regulations* is paramount when generating and utilizing synthetic medical data to prevent unauthorized disclosure of sensitive patient information. In response to these challenges, researchers and developers are exploring innovative approaches to synthetic data generation that prioritize patient privacy and comply with regulatory requirements. Techniques such as *federated learning (FL)* [28], *differential privacy* [37], and *data anonymization* [32] are being investigated to mitigate privacy risks while enabling the development of robust generative AI models.

In particular, the FL-based FedSyn method [2] allows AI models to be trained across multiple decentralized synthetic data sources without directly accessing raw patient data. Each participating healthcare institution retains control over its data, ensuring compliance with medical privacy regulations like HIPAA and GDPR. By aggregating model updates rather than raw data, FedSyn can enable the training of robust generative AI models while preserving privacy and copyright regulations. *Differential privacy* techniques add noise to individual data points before they are used for training generative AI models. This ensures that no single patient's data can be inferred from the model's outputs, thereby protecting medical privacy and copyright regulations. *De-identifying and Anonymizing medical images* by removing or obfuscating identifying information such as patient names, dates of birth, medical record numbers, etc., can help protect patient privacy while enabling the use of synthetic data to train generative AI models. Additionally, strategies such as *consent for medical data* and *intellectual property* should be examined when developing and deploying generative AI models in medical imaging.

5   Measuring Fairness, Bias, and Diversity

Fairness metrics assess whether the generative AI model produces *unbiased results* across different patient demographics, such as age, gender, ethnicity, or socioeconomic status. Equal opportunity is a critical aspect of fairness, which entails ensuring that the generative AI model provides equitable access to accurate medical image generation for all patients, regardless of their demographic characteristics [31]. This means the model should not exhibit realism and correctness based on factors like race or gender. On the other hand, procedural fairness ensures that the processes involved in medical data collection, model training, and decision-making are *transparent, accountable, and unbiased*. This includes transparent documentation of medical data sources, model architectures, and evaluation metrics, as well as mechanisms for addressing and mitigating biases at each stage of the AI pipeline [12].

Bias assessment involves identifying and mitigating biases in the generative AI model's training data or algorithms that may lead to skewed or inaccurate image generation [8]. Firstly, *biases may originate in the datasets* that are used to train AI models and might inadvertently reflect societal prejudices or inequalities present in the data sources. For instance, historical biases in medical records or imaging datasets can lead to skewed representations of certain demographic groups, potentially resulting in biased AI predictions. Secondly, *biases can be introduced during the algorithm design*, where choices made in AI model architecture, feature selection, or optimization techniques may perpetuate or amplify existing biases in the data. Lastly, *biases can also emerge from user interactions* with generative AI systems, where feedback loops or reinforcement learning mechanisms may reinforce and exacerbate biases present in the system's outputs. Therefore, it is imperative to identify, mitigate, and address biases at each stage of the machine learning pipeline to ensure the development of fair, transparent, and unbiased AI systems.

Diversity of medical image generation encompasses the ability of the AI system to generate images that represent a *wide range of clinical scenarios, anatomical structures, imaging modalities, and disease presentations*, ensuring that the AI system can effectively capture the heterogeneity present in clinical practice. It ensures the AI system's *robustness, generalizability, and clinical relevance*. Additionally, diversity of generation enables AI models to adapt to diverse patient populations and imaging settings, facilitating their broader applicability and adoption in healthcare settings. To evaluate and promote diversity in generative AI outcomes for medical imaging, researchers may employ metrics based on the entropy of neural network encodings [16]. Overall, generation diversity ensures that the AI system can effectively capture the complexity and heterogeneity of real-world clinical data, thereby enhancing its utility and impact in medical imaging applications.

6   Conclusions

This study highlights the ethical complexities of deploying generative AI in medical imaging. Our study emphasizes fairness, patient privacy, accountability, algorithmic and data bias, and transparency, emphasizing the need for the responsible use of generative AI in medicine. We have identified practical and ethical challenges, from data realism and generalizability to transparency, validation, and regulation, emphasizing the need for continuous scrutiny and improvement. Additionally, with some qualitative examples, we showed the effect of adversarial perturbations on medical image segmentation performance and the effectiveness of diffusion models in generating realistic medical images. Moreover, we also provide some standard regulations on synthetic medical data and discuss ways to ensure fairness and mitigate bias problems and diversity challenges. We advocate for collaboration between academic researchers, ethicists, clinicians, and regulatory bodies to ensure ethical considerations are met for developing responsible generative AI in medicine.

Acknowledgements. This project is supported by NIH funding: R01-CA246704, R01-CA240639, U01-DK127384-02S1, and U01-CA268808.

Conflicts of Interest. The authors have no relevant financial or non-financial interests to disclose.

References

1. Amodei, D., Olah, C., Steinhardt, J., Christiano, P., Schulman, J., Mané, D.: Concrete problems in AI safety. arXiv preprint arXiv:1606.06565 (2016)
2. Behera, M.R., Upadhyay, S., Shetty, S., Priyadarshini, S., Patel, P., Lee, K.F.: FedSyn: synthetic data generation using federated learning. arXiv preprint arXiv:2203.05931 (2022)
3. Cardoso, M.J., et al.: MONAI: an open-source framework for deep learning in healthcare. arXiv preprint arXiv:2211.02701 (2022)
4. Carlini, N., et al.: Extracting training data from diffusion models. In: Proceedings of the 32nd USENIX Security Symposium (USENIX Security 23), pp. 5253–5270 (2023)
5. Chen, M., et al.: Evaluating large language models trained on code. arXiv preprint arXiv:2107.03374 (2021)

6. Cheng, H.T., Thoppilan, R.: LaMDA: towards safe, grounded, and high-quality dialog models for everything. Google AI Blog (2022)
7. Chowdhery, A., et al.: Palm: scaling language modeling with pathways. J. Mach. Learn. Res. **24**(240), 1–113 (2023)
8. Ferrara, E.: Fairness and bias in artificial intelligence: a brief survey of sources, impacts, and mitigation strategies. Sci **6**(1), 3 (2023)
9. Gerke, S., Minssen, T., Cohen, G.: Ethical and legal challenges of artificial intelligence-driven healthcare. In: Artificial intelligence in healthcare (2020)
10. Goodfellow, I., et al.: Generative adversarial networks. Commun. ACM **63**(11), 139–144 (2020)
11. Goodfellow, I.J., Shlens, J., Szegedy, C.: Explaining and harnessing adversarial examples (2015). https://arxiv.org/abs/1412.6572
12. Grote, T., Keeling, G.: On algorithmic fairness in medical practice. Camb. Q. Healthc. Ethics **31**(1), 83–94 (2022)
13. Guibas, J.T., Virdi, T.S., Li, P.S.: Synthetic medical images from dual generative adversarial networks. arXiv preprint arXiv:1709.01872 (2017)
14. Hao, S., Han, W., Jiang, T., Li, Y., Wu, H., Zhong, C., Zhou, Z., Tang, H.: Synthetic data in AI: challenges, applications, and ethical implications. arXiv preprint arXiv:2401.01629 (2024)
15. Ho, J., Jain, A., Abbeel, P.: Denoising diffusion probabilistic models. Adv. Neural. Inf. Process. Syst. **33**, 6840–6851 (2020)
16. Ibarrola, F., Grace, K.: Measuring diversity in co-creative image generation. arXiv preprint arXiv:2403.13826 (2024)
17. Kingma, D.P., Ba, J.: Adam: a method for stochastic optimization (2017). https://arxiv.org/abs/1412.6980
18. Kingma, D.P., Welling, M.: Auto-encoding variational bayes. arXiv preprint arXiv:1312.6114 (2013)
19. Le Scao, T., et al.: Bloom: a 176b-parameter open-access multilingual language model (2022)
20. Li, Y., et al.: Competition-level code generation with alphacode. Science **378**(6624), 1092–1097 (2022)
21. Litjens, G., et al.: A survey on deep learning in medical image analysis. Med. Image Anal. **42**, 60–88 (2017)
22. Liu, X., Song, L., Liu, S., Zhang, Y.: A review of deep-learning-based medical image segmentation methods. Sustainability **13**(3), 1224 (2021)
23. Liu, Z., et al.: RadImageGAN–A Multi-modal Dataset-Scale Generative AI for Medical Imaging. arXiv preprint arXiv:2312.05953 (2023)
24. Midjourney: Midjourney home. https://www.midjourney.com/home (2023). Accessed 24 Sep 2023
25. Musalamadugu, T.S., Kannan, H.: Generative AI for medical imaging analysis and applications. Future Med. AI FMAI5 **1**(2) (2023)
26. OpenAI: Gpt-4. https://openai.com/research/gpt-4 (2023). Accessed 25 March 2024
27. Paszke, A., et al.: PyTorch: an imperative style, high-performance deep learning library. Adv. Neural Inf. Process. Syst. **32** (2019)
28. Rauniyar, A., et al.: Federated learning for medical applications: a taxonomy, current trends, challenges, and future research directions. IEEE IoT J. (2023)
29. Rombach, R., Blattmann, A., Lorenz, D., Esser, P., Ommer, B.: High-resolution image synthesis with latent diffusion models. In: Proceedings of the IEEE/CVF Conference on Computer Vision and Pattern Recognition, pp. 10684–10695 (2022)

30. Ronneberger, O., Fischer, P., Brox, T.: U-net: Convolutional networks for biomedical image segmentation. In: Medical Image Computing and Computer-assisted Intervention–MICCAI 2015: 18th International Conference, Munich, Germany, October 5-9, 2015, proceedings, part III 18, pp. 234–241 (2015)
31. Shen, A., Han, X., Cohn, T., Baldwin, T., Frermann, L.: Optimising equal opportunity fairness in model training. arXiv preprint arXiv:2205.02393 (2022)
32. Shin, H.C., et al.: Medical image synthesis for data augmentation and anonymization using generative adversarial networks. In: Third International Workshop, SASHIMI, pp. 1–11 (2018)
33. Stability AI: Stable diffusion SDXL-1 announcement. https://stability.ai/news/stable-diffusion-sdxl-1-announcement (2023). Accessed 24 Sep 2023
34. Touvron, H., et al.: LLaMA: open and efficient foundation language models. arXiv preprint arXiv:2302.13971 (2023)
35. Uzunova, H., Schultz, S., Handels, H., Ehrhardt, J.: Unsupervised pathology detection in medical images using conditional variational autoencoders. Int. J. Comput. Assist. Radiol. Surg. **14**, 451–461 (2019)
36. Voigt, P., Von dem Bussche, A.: The General Data Protection Regulation (GDPR). Springer International Publishing **10**(3152676), 10–5555 (2017)
37. Xin, B., et al.: Federated synthetic data generation with differential privacy. Neurocomputing **468**, 1–10 (2022)
38. Yu, M., et al.: How good are synthetic medical images? An empirical study with lung ultrasound. In: International Workshop on Simulation and Synthesis in Medical Imaging, pp. 75–85 (2023)

Author Index

E. Puyol-Ant et al. (Eds.): FAIMI 2024/EPIMI 2024, LNCS 15198, pp. 189–190, 2025.
https://doi.org/10.1007/978-3-031-72787-0